# Making the Most of Your Catch

*An Angler's Guide*

# Making the Most of Your Catch

*An Angler's Guide*

Ian Dore

An Osprey Book
Published by Van Nostrand Reinhold
New York

An Osprey Book
(Osprey is an imprint of Van Nostrand Reinhold)

Drawings by Lisa Murrin

Copyright © 1990 by Van Nostrand Reinhold

Library of Congress Catalog Number: 89-70669
ISBN: 0-442-00195-9

All rights reserved. No part of this work covered by the copyright hereon may be reproduced or used in any form by any means—graphic, electronic, or mechanical, including photocopying, recording, taping, or information storage and retrieval systems—without written permission of the publisher.

Printed in the United States of America.

Van Nostrand Reinhold
115 Fifth Avenue
New York, New York 10003

Van Nostrand Reinhold International Company Limited
11 New Fetter Lane
London EC4P 4EE, England

Van Nostrand Reinhold
480 La Trobe Street
Melbourne, Victoria 3000, Australia

Nelson Canada
1120 Birchmount Road
Scarborough, Ontario M1K 5G4, Canada

16 15 14 13 12 11 10 9 8 7 6 5 4 3 2 1

**Library of Congress Cataloging-in-Publication Data**

Dore, Ian, 1941-
   Making the most of your catch: an angler's guide / Ian Dore
     p. cm.
   Includes bibliographical references (p. ).
   ISBN 0-442-00195-9     $14.95
   1. Fish as food—Preservation. 2. Cookery (Fish) I. Title.
   TX612.F5D57     1990
   641.4'94—dc20                   89-70669
                                          CIP

# Contents

*Preface*   ix

*Chapter One*   Handling and Processing Fresh Fish   1

    Handling Live Fish   1
    Gutting and Bleeding   2
    How to Clean Fish   3
    Chilling   6
    Alternative Ways to Chill   7
    Salt and Burlap   9
    Processing at Sea   10
    Processing   11
    Fresh Fish Storage at Home   16
    Is Your Fish Good?   18
    Deterioration in Storage   18

*Chapter Two*   Freezing the Catch   22

    Condition of the Fish   22
    Freezing Rate   23
    Glazing   25
    Packaging   26
    Labeling   29
    Storage Time   30
    Thawing   33

|  |  |
|---|---|
| Refreezing | 33 |
| Summary | 34 |

## Chapter Three  Salting Fish — 35

| | |
|---|---|
| How Salt Preserves | 36 |
| Preparing the Fish | 37 |
| Salt | 39 |
| Pickle Curing | 40 |
| How to Salt Fish | 41 |
| Some Helpful Hints for Salting Fish | 46 |

## Chapter Four  Pickling and Marinating Fish — 49

| | |
|---|---|
| Ingredients for Pickling and Marinating | 50 |
| Recipes | 51 |

## Chapter Five  Smoking Fish — 58

| | |
|---|---|
| Equipment for Smoking Fish | 58 |
| Preparation | 60 |
| The Smoking Process | 61 |
| After Smoking | 67 |
| Liquid Smoke | 68 |
| Cod Roe | 68 |

## Chapter Six  One Man's Trash is Another's Treasure — 69

| | |
|---|---|
| Whiting | 69 |
| Red Hake (Mud Hake, Squirrel Hake) | 71 |
| Dogfish | 71 |
| Squid (Calamari) | 76 |
| Skates and Rays | 81 |

| | |
|---|---:|
| Anglerfish (Monkfish, Allmouth, Goosefish, Ocean Blowfish, Mother-in-law Fish) | 84 |
| Sea Robins (Sculpins) | 86 |
| Et Cetera | 86 |
| Regional Fishes | 89 |

## *Chapter Seven*   Yes, Virginia, Seafood is Safe to Eat — 96

| | |
|---|---:|
| Poisoning Due to Biotoxins | 97 |
| Poisoning Due to Chemicals | 103 |
| Parasitic Infections | 105 |
| Bacterial Food Infections or Intoxicants | 109 |
| Virus Infections | 112 |
| Radioactivity | 112 |
| Allergies | 113 |
| Bites and Stings | 113 |
| Bacterial Skin Infections | 114 |

## *Chapter Eight*   Fish and Nutrition — 116

| | |
|---|---:|
| Nutrient Sources | 117 |
| Proteins and Fats | 117 |
| Vitamins and Minerals | 123 |

## *Chapter Nine*   Basic Techniques for Cooking Fish — 125

| | |
|---|---:|
| The Canadian Cooking Rule | 125 |
| Baking | 126 |
| Oven-frying | 127 |
| Broiling | 128 |
| Charcoal Broiling | 128 |
| Frying | 129 |

| | |
|---|---|
| Deep-fat Frying | 129 |
| Pan-frying (Sautéing) | 130 |
| Stir-frying | 131 |
| Poaching and Steaming | 132 |
| Braising and Stewing | 133 |
| Microwave Cooking | 134 |
| Wines in Fish Cookery | 135 |
| Using Leftover Fish | 135 |
| Cookbooks | 136 |

Glossary — 137

Further Reading — 149

Index — 157

# Preface

This is a comprehensive guide for recreational fishermen on how to care for and use the fish they catch. It covers handling the catch so that it arrives home in good condition; preparing and processing it; freezing, salting, smoking and pickling it. The book also covers the nutritional benefits and safety aspects of eating fish.

There are an estimated 17 million marine anglers in the U.S.A. who catch 20 percent of the country's edible finfish. There are probably even more freshwater fishermen. Many anglers do not handle their fish properly, so their catch ends up in the garbage can. Some even throw back much of what they catch! With the price of fish high and increasing, every sport fisherman can benefit from using his catch to the fullest extent. The fish you put on the table saves money at the store. It can also be fresher and more varied than store-bought fish.

The book is based on material assembled by scientists at the Laboratory of the National Marine Fisheries Service in Gloucester, Massachusetts, who together have more than 400 years of professional experience in fishery technology as well as a universal passion for angling. The Laboratory has been carrying out research on the use of fish for food for 35 years and is nationally and internationally recognized for its work on processing and preservation of fish and shellfish. Among the many people who provided substantial material for this book are the following:

**Robert J. Learson**, Laboratory Director, with the chapters "Handling and Processing Fresh Fish" and "One Man's Trash is Another's Treasure." Bob has been enormously helpful in checking and rechecking everything, answering questions and in providing encouragement over a period of several years so that we eventually finished the job and got into print. Without him, this book would definitely not exist.

**Joseph J. Licciardello**, Research Food Technologist, with the chapters on freezing and seafood safety.

**John D. Kaylor**, Research Food Technologist, with the chapters on salting, pickling and on regional fishes.

**Joseph H. Carver**, Research Chemist, with the material on smoking.

**Judith Krznowek**, Research Chemist, with the nutrition chapter.

**Betty Tukhunen**, Food and Nutrition Scientist, with the fish cookery section.

Another person who applied his immense expertise to the project was Richard Stroud. Dick is a former president of the American Fisheries Society, a former chairman of the Natural Resources Council of America and was also the chief executive officer of the Sport Fishing Institute. His knowledge of fish and fishing was vital and his advice always to the point and valuable.

Amy Spinthourakis is a most able researcher and copy editor. She has cheerfully corrected countless passive sentences and other imperfections frowned upon by Strunk and White.

I thank these people and all the others who made this book possible. Remaining imperfections and errors are mine, not theirs.

*Ian Dore*
*New York, December 1989*

Chapter One

# Handling and Processing Fresh Fish

Millions of people enjoy the pastime of fishing. To many, it is just a sport. To others, it is also food on the table. Fish can be expensive. By using the fish you catch, you help to defray the cost of fishing. Many fishermen do not know how to handle their catch properly; therefore, they can enjoy their fish only for a short time. There are still others who do not know that there are many ways of preserving their fish so that they can enjoy it months later. Learning to handle your catch properly takes just a little practice, but the benefits are long lasting and valuable.

## Handling Live Fish

One of the enjoyments of rod-and-reel fishing is the sensation of catching a fish that fights back. However, many fishermen do not care for heavy tackle that allows "horsing them in." When a fish struggles violently for a time, it uses up much of its body sugar. As a result, some of the natural sweetness is lost from the flesh. The best compromise for enjoying the fishing as well as the fish is to stun or kill the fish as soon as you land it. Never allow the fish to flop around in the boat or, worse, to struggle for hours in a bucket or on a stringer dangling in the water.

If you use a gaff, always strike the fish in the head. If you gaff it in the side, blood and bacteria from the gut will penetrate the muscle slab—the edible fillet—and increase the chance of spoilage.

## Gutting and Bleeding

Gut and bleed all fish as soon as possible after landing. Enzymes that are present in the flesh, stomach and blood of the fish cause chemical changes that are necessary for life. After the fish dies, its digestive and blood systems cease to function. These systems then become unbalanced and can cause rapid deterioration of the flesh. The gut enzymes migrate into the flesh where they continue to perform their natural function of breaking down tissue: the flesh softens, which allows bacteria to grow as the fish partially digests itself. This happens most strongly when fish were feeding before being caught, because their stomachs then contain large quantities of the powerful digestive enzymes. Ungutted fish, therefore, can soften and spoil rapidly. Further, fish that are adapted to living in cold water have enzymes that work well even at refrigerated temperatures. This includes all the species generally caught off both northern Atlantic and northern Pacific coasts, as well as in the larger rivers, lakes and reservoirs.

Spoilage bacteria are usually confined to the surface slime, gills and intestines of the fish. Bacterial action, aided by the changes induced by enzymes, is the main cause of fish spoilage. Although spoilage bacteria are few in number on fish that are alive, spoilage bacteria can and do multiply rapidly after the fish dies. These organisms, often reinforced by additional bacteria introduced from unsanitary handling or dirty equipment, increase in numbers. Eventually, these bacteria penetrate the skin and belly wall and pass into the flesh.

**Figure 1.** Gutting: ripping to vent.

Thorough bleeding is of primary importance. Blood coagulating in the fillet causes discoloration. The blood also contains trace amounts of metals and enzymes that can promote spoilage and rancidity even during frozen storage. Cleaning the fish immediately after it is caught ensures that it is properly bled.

## How to Clean Fish

Cleaning, gutting and eviscerating all mean removing the guts from the fish. The process, whatever you want to call it, is vitally important.

Commercial fishermen commonly gut a fish by holding it by the head, inserting the knife between the nape bones and ripping the belly open to the vent (see Figure 1). Other fishermen grasp the fish by the back and, starting at the vent, rip upwards to the nape. Whichever way you do it, make sure that the belly cavity is fully opened. Remove the entire contents of the cavity, including the soft, dark area along the backbone, which is the kidney. Protecting the kidney is a membrane, which must be cut first. Then, after cutting the membrane, scrape out the kidney material with the knife (see Figure 2).

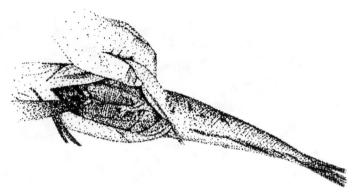

**Figure 2.** Removing (scraping) kidney.

**Figure 3.** Removing gills.

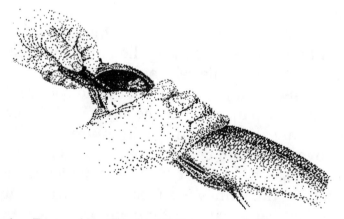

**Figure 4.** Removing gills (cont'd).

You should also remove the gills, especially in hot weather. To do this, cut between the nape bones and remove the gills with a sharp pull or by cutting them free with a knife (see Figures 3 and 4).

Wash the fish after gutting and bleeding to eliminate all remnants of blood, guts and slime. Rinsing in a solution of one tablespoon of chlorine bleach in four gallons of water ensures maximum cleansing.

When the fish are really biting, it is too time-consuming to gut and bleed each fish as it is caught. Simply cut the fish quickly across the throat and notch the tail to kill

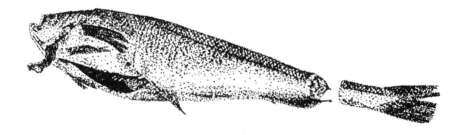

Figure 5.   Bleeding: tail cut.

Figure 6.   Bleeding: throat cut.

and bleed them (see Figures 5 and 6). Then hold the fish temporarily in a bucket of cold saltwater solution (brine) or seawater for gutting later when the fishing quiets down.

## Chilling

Keeping the fish cool is by far the most important factor in maintaining high eating quality.

Chilling does not stop spoilage completely, but it does slow it down. Even a few degrees can be critical. Gutted fish held at 32°F (the temperature of melting ice) last twice as long as those held at 37°F. Bacteria, which are the main cause of fish spoilage, can still multiply at cold temperatures. Their growth rate approximately doubles for each additional 10°F in temperature. If you leave your fish on the hot deck of the boat, or dangle your catch on stringers in 60°F or 70°F water, or, worst of all, put the fish without ice in the trunk of your car on a hot day, you will be in for a most unpleasant surprise when you get the fish home.

Ice is the most effective and convenient method of chilling fish. The type of ice is important, as well as the quantity. If possible, use crushed ice or flake ice rather than blocks or cubes. Blocks and cubes have sharp edges and corners that may penetrate the fish, leaving cuts and tears where bacteria are more likely to grow. Ice absorbs heat from its surroundings as it melts. A large block of ice does not melt as quickly as flake ice and therefore absorbs heat more slowly, taking longer to chill the fish. Although flake ice and crushed ice are preferable, blocks and cubes are far better than no ice. Use the best that is available.

Use one pound of ice to three pounds of fish. Preferably, store the fish with ice all around them. The gut cavities of larger fish should also be packed with ice. Place the fish bellies down to allow the melting ice to drain. When storing a large number of fish, especially a large number of small fish, mix ice with each layer to ensure maximum chilling.

The best container for icing is one that has holes in the bottom or a spigot to allow the meltwater to drain. The water from melting ice not only chills but also washes blood and bacteria from the fish. Good drainage is necessary so that the fish do not lie in contaminated meltwater. Insulated containers or buckets without drains can easily be fitted with false bottoms that will hold the fish above the meltwater.

Properly gutted and bled fish can be held in ice for four or five days with little loss of eating quality and nutritional value.

## Alternative Ways to Chill

Other methods of chilling commonly used by the commercial fishing industry can also be used by sport fishermen. On larger commercial boats that have refrigeration systems, fish is sometimes stored in refrigerated seawater or saltwater. This can be a satisfactory alternative to icing. Chilled seawater systems are simply insulated tanks with cooling coils from a refrigeration compressor. About 3 percent of the weight of seawater is salt, which lowers its freezing point to about 28°F. Some of the advantages of refrigerated seawater are:

1. It eliminates the inconvenience of carrying ice.

2. The fish chill rapidly when immersed in 28°F brine or seawater, so this helps to maintain the very best quality.

3. There is less physical damage to the fish when they float in brine than when they are packed in ice.

Refrigerated seawater systems in the commercial fishing industry are generally preferred for herring, mackerel and menhaden because so many are caught at one time that careful handling and proper icing are impossible. For sport as well as commercial fishermen, refrigerated brine or seawater is especially good for softer fleshed fish such as shad, crappies, pickerel, sea trout, whiting, red hake and squid. Refrigerated seawater is not restricted only to saltwater fishermen. Freshwater fishermen can make brine that is equivalent to seawater by mixing 28 pounds of salt with each 100 gallons of clean, fresh water. To make smaller quantities (for when the fish aren't biting), the basic formula is 1.4 pounds (22½ ounces) of salt for 5 gallons of water.

The recommended ratio of fish to brine is three or four pounds of fish to one pound (one pint) of brine or seawater. Fish held in refrigerated brine for two or three days will look and smell as if they were just caught.

Slush ice (or chilled brine) is another chilling method that has some advantages over regular icing, especially for softer fleshed species. Chilled seawater is a mixture of clean seawater and crushed or flake ice. Chilled brine is a mixture of clean water, 3 percent salt and the ice. When you are at sea, add just enough clean seawater or brine to the tank of ice to make "slush" immediately before fishing. Make sure to gut and clean the fish as you catch them, and immediately add them to the chilled mixture. The relative volumes of ice and water depend on the air and water temperatures. Obviously, you will need more ice in the warmer months. A mixture of eight pounds of ice to each gallon of water is a reasonable starting mixture. This should hold about 25 pounds of fish in a well-insulated container for a couple of days. Make sure there is a sufficient amount of ice to cool the brine to 32°F and to keep it at that temperature for the duration of your trip. The amount of ice required to cool seawater to 32°F from several different water temperatures is as follows.

40°F: add 4.7 pounds of ice for each 10 gallons of water
50°F: add 10.5 pounds of ice for each 10 gallons of water
60°F: add 16.4 pounds of ice for each 10 gallons of water
70°F: add 22.2 pounds of ice for each 10 gallons of water

Remember to refresh the ice mixture frequently with more ice to keep the temperature as close as possible to 32°F. The amount of ice depends on three factors:

1. the insulating values of the holding tank,
2. the holding time, and
3. how much fish is added.

As a rough guide, even a well-insulated tank requires about seven times the quantities of ice indicated above to maintain chilling temperatures for 60 hours.

As mentioned before, it is very important when using both refrigerated and chilled seawater systems that the fish is properly gutted and bled before being chilled. Any remnants of viscera or blood will contaminate the water and diminish the eating quality of all the fish.

## Salt and Burlap

Ice may not always be readily available out at sea, and it is seldom practicable for beach fishermen to pack heavy loads of ice and carry it several miles down the beach. If this is the case, an alternative is to carry some salt and several pieces of burlap. After cleaning the fish, rub salt into the belly cavity, using about one tablespoon of salt to each pound of fish. Salt the skin lightly as well. Put the fish on a pad of dampened burlap or, if available, wet seaweed, and store the fish in a basket or box. Cover the box or basket with several layers of wet burlap, leaving an airspace of a few inches between the cloth and the fish. Keep the burlap damp and shade the box from the sun. On a dry day, evaporation from the burlap keeps the fish cool. At the least, the use of salt has some preservative effect. Note that in hot and humid weather there is hardly any evaporation. Without evaporation, there is little or no cooling.

Burying the fish in the sand near the water line is a reasonably good chilling method, especially at the seaside. Burial prevents the fish from drying out, and the moist sand a few inches below the surface is usually several degrees cooler than the air temperature. Mark the location well and watch the incoming tide carefully.

## Processing at Sea

There are advantages and disadvantages to processing the fish into edible pieces while you are still at sea. On the positive side, cutting fish into fillets or steaks reduces by about half the volume of material that you have to ice and transport. If you are planning to freeze the catch, cutting the fish is best done immediately before you put it in the freezer. On the negative side, conditions on a fishing trip are rarely suitable for careful work.

*Rigor mortis*, the process of "death stiffening," can affect the eating quality if you plan on freezing the fish. Bacterial growth in fish that are in rigor is slow: keeping the fish in rigor for as long as possible helps to ensure top quality. Rigor is extended by minimizing the struggling of the fish before it dies and by chilling it as close as possible to 32°F immediately after catching it. In general, it is better to fillet the fish at home, after it has passed through rigor. (There is a detailed discussion of rigor and its effects on fish in Chapter Two, Freezing the Catch.) However, you must gut the fish immediately after catching it to preserve the best possible quality.

Perhaps the best compromise at sea is to cut off the heads of the fish, saving about one-third of the total weight. Although the heads are useful "handles" when you fillet the fish and there are edible cheeks and tongues in larger fish heads (see Chapter Six), the loss of these delicacies may not outweigh the savings in weight and space.

## Processing

*Scaling.* Any fish that are to be prepared for the table with the skin on should be scaled. Remove scales by scraping the sides of the fish with a relatively dull knife, starting from the tail and working towards the head, that is, against the grain. Wash away any loose scales.

*Filleting.* Filleting fish is not difficult. With a little practice you can soon learn to produce smoothly cut pieces that yield as much flesh as possible from the carcass. A fillet is the side muscle from the fish, removed from the

**Figure 7.** Filleting knives.

> **Box 1: Filleting Board**
>
> Commercial filleters in New England and Canada have been using a comparatively new style of cutting board because it helps to increase the fillet yield from their fish. The design of this board is based on industrial engineering studies carried out in Newfoundland. It differs from the traditional cutting board by being angled towards the cutter at a slope of 8 degrees (or one inch in height for a 12-inch-wide
>
>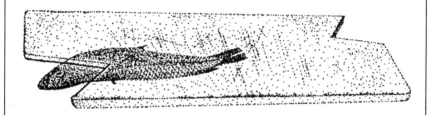
>
> board) (see Figure). It also has two slots for placing the head of the fish while you cut the second fillet. One slot is for large fish, the other for small fish. The 8-degree slope allows the knife angle to be more natural; the head slots give the cutter a firmer base for filleting the second side.

backbone. It may have skin left on or removed. A fillet may also contain small pin bones, or they may be cut out.

A good fillet knife is absolutely necessary. Professional fish cutters use a thin-blade knife similar to the one in the illustrations. The blade should be sharp but never razor sharp: a blade that is too sharp will cut into the bones rather than just glide over them. Suitable knives are illustrated in Figure 7. A step-by-step filleting operation of a round fish is illustrated in Figures 8, 9 and 10.

Box 1 describes a new type of cutting board that helps increase fillet yield.

*Skinning.* Fillets from fish that have a relatively thick skin and firm flesh are usually skinned before use. Such fish include saltwater species like cod, pollock, flounder, Pacific rockfish, ocean catfish and cusk, as well as freshwater species like bass, walleye and channel catfish. Other species such as haddock, whiting, mackerel and hake are usually not skinned. These species have relatively soft flesh, and leaving the skin on helps to hold the meat together during cooking. Skinning fillets is shown in Figure 11. Avoid using a razor-sharp knife as this will cut

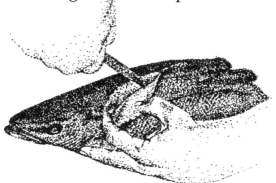

Figure 8. Filleting: start cutting behind the head.

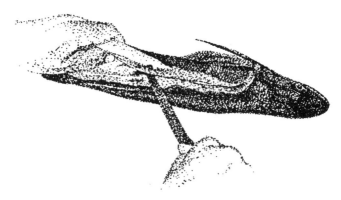

Figure 9. Filleting (cont'd).

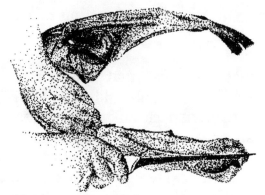

Figure 10.  Scraping viscera from fillet.

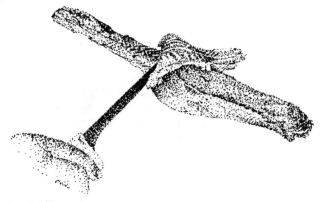

Figure 11.  Fillet: skinning.

through the skin rather than follow the plane of the skin. To stop the fillet from slipping and to help make the initial cut, use a fork or a short piece of wood with a bottle cap nailed to the end to hold the tail end of the fillet.

*Steaking.*  Big fish are commonly cut into steaks. Since larger fish have larger backbones, you need two knives for steaking. Use a fillet knife for cutting the flesh and a much heavier knife or cleaver to cut through the backbone. The tail section, where the fish is too narrow for proper steaking, should be filleted. Leave the skin on

Figure 12. Steaking: severed backbone.

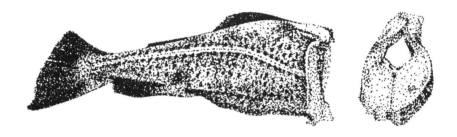

Figure 13. Removing steak.

steaks to hold the flesh together during cooking (see Figures 12 and 13).

Head and dress fish that are too small for filleting. This means cutting off the head, fins and tail. Some small fish can be gutted and headed at the same time (a process known in the industry as "nobbing"): cut the head from the back, pulling out the guts still attached to the head. Small fish can also be split or butterflied for easy removal of the bones after cooking. To produce a butterfly fillet, cut the fish through from the back along the backbone, leaving the fish joined at the belly side. Open it up, with the backbone still attached to one side. The backbone can be removed to produce a fillet, or left on if the fish is, for example, a small haddock to be used for making a "Finnan" (see Chapter Five for details of smoking your catch).

There is an excellent method for dressing small flounder that gives a much greater meat yield than filleting. Cut behind the head of the flounder. Trim off any remaining bones from the side "frill," or fin, and cut off the tail. Use pliers to rip off the dark skin. The result is a dressed flounder that is excellent for broiling, baking or frying. The process is shown in Figures 14 to 17.

## Fresh Fish Storage at Home

Temperature control is critical to maintain maximum food value and flavor. Spoilage bacteria grow very well on fish at relatively low temperatures. Changes in temperature of only a few degrees can make a big difference to the eating quality in as little as one or two days.

Before putting any fish in the refrigerator, wash it in clean tap water. Scientific studies have shown that this simple procedure can remove almost 90 percent of surface bacteria.

Figure 14.   Dressing flounder: first cut.

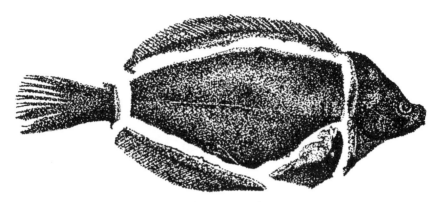

Figure 15. Dressing flounder: trimming.

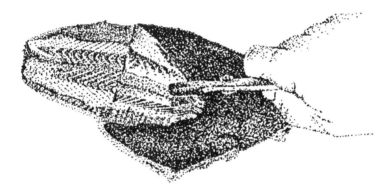

Figure 16. Dressing flounder: skinning dark side.

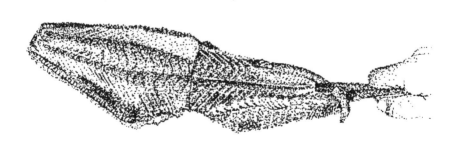

Figure 17. Dressing flounder: skinning dark side.

## Is Your Fish Good?

There is no processing or preservation method that improves the initial quality of the raw material. Odor is the best way to tell how good the fish is, but there are many other points to observe. Properly handled, fresh fish have clear, bulging eyes, relatively odorless body cavities, clean, shiny skins and firm flesh. Mistreated fish have filmy and sunken eyes, body cavities that smell fishy and soft flesh.

A quick method of checking the firmness of your fish: Make an indentation in the side of the fish with your thumb. In good, fresh fish, the indentation soon disappears; in a poor quality fish, the indentation remains. Fillets or steaks from fresh fish appear translucent and glossy. They have very little odor or just a slight, briny, seaweed-like odor. The flesh from poorer quality fish is more opaque and less glossy. The odor is slightly fishy or sour.

Table 1 is adapted from a checklist used by commercial fish plants to evaluate the fish they receive. You may find this useful in evaluating your own fish. Remember that odor is the major determinant. First rinse the fish to remove accumulated gases and odors from storage, packaging or meltwater. Then smell it. If it smells bad, it certainly is bad.

## Deterioration in Storage

Table 2 describes appearance and odor changes in cod stored for 10 days at 33°F, starting from freshly caught fish held with strict temperature control. It is still in pretty good condition after one week. Figure 18 shows how rapidly fish deteriorates when storage temperatures are higher: at 46°F, the same fish would be in rather poor shape after less than four days. If temperature control is

Table 1: Fresh Fish Quality Assessment

| | Top of catch | Good | Fair | Poor |
|---|---|---|---|---|
| ODOR | Fresh, strong, seaweedy shellfishy | No odor; neutral odor | Definite musty, mousy, bready, malty odor *Process immediately* | Acetic, fruity, sulphic, faecal REJECT |
| GUT CAVITY | Glossy, brilliant, difficult to tear from flesh | Slightly dull, difficult to tear from flesh | Somewhat gritty, somewhat easy to tear from flesh | Gritty, easily torn from flesh |
| GILLS | Bright red, mucus translucent | Pink, mucus slightly opaque | Gray, bleached, mucus opaque and thick | Brown, bleached, mucus yellowish gray, clotted |
| EYES | Convex, black pupil; translucent cornea | Flat, slightly opaque pupil | Slightly concave, gray pupil; opaque cornea | Completely sunken gray pupil; opaque, discolored cornea |
| OUTER SLIME | Transparent or water white | Milky | Yellowish gray, some clotting | Yellowish brown, very clotted and thick |
| SKIN | Bright, shining, irridescent, no bleaching | Wavy, slight dullness, slight loss of brightness | Dull, some bleaching | Dull, gritty, marked bleaching and shrinkage |

Source: Paquette, Gerald, 1983, *Fish Quality Improvement*. Van Nostrand Reinhold/Osprey Books.

poor, bacteria grow quickly and spoilage accelerates. Compared with fillets stored at 33°F, those kept at 42°F spoil twice as fast and those stored at 46°F spoil three times as fast (see Figure 18). Home refrigerators are often set at 40°F, sometimes higher. This is too high for keeping your fish fresh for a few days.

Table 2: Changes in Cod Fillet Quality Stored at 33°F

| No. of Days in Storage | Appearance | Odor |
|---|---|---|
| 1-4 | Glossy<br>Translucent | Neutral<br>Briny<br>Seaweedy |
| 5-6 | Little or no glossiness | Very slight fishy<br>Stale or musty |
| 7-8 | Dull<br>Opaque<br>Very slight yellow or tan | Slight fishy<br>Sour, acidic |
| 9-10 | Opaque<br>Yellowing | Definite fishy<br>Sour or acidic<br>Persistent ammonia |

The best way to keep fish fresh at home is to have a second refrigerator set close to freezing. If this is not possible, use ice. First, pack the fish in sealed plastic bags, excluding as much air as possible. Put the bags in ice-filled trays in the refrigerator. The trays do not need to be covered. Empty the meltwater from the trays periodically, and add more ice to replace what melted.

The vegetable or meat bins in the refrigerator can be used for this if they are leak-proof. This method will keep the fish chilled to 33°F for three or four days without much loss of quality.

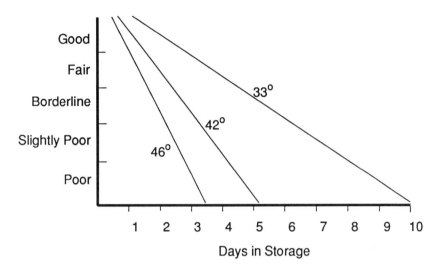

**Figure 18.** Effect of storage temperature on fresh quality of fish fillets. (Source: National Marine Fisheries Service.)

If you have fish you want to keep for longer periods, it must be frozen or preserved in some other way. The following chapters tell you how.

*Chapter Two*

# Freezing the Catch

It was a great fishing trip and you came home with a lot of fish. You ate some, gave more to the neighbors and packed the rest in the refrigerator. There is no way that you can eat it all before it spoils. Ways to preserve part of your catch are explained in the next few chapters.

Freezing is probably the easiest and most popular way to preserve food at home and is particularly effective for seafoods. It is not only easy to freeze fish; it is also easy to prepare it later for eating. Usually, you do not even need to thaw fish before using it. You can cook fish while it is still frozen without any problem, so that using your frozen fish is very simple.

## Condition of the Fish

It is important that the fish be as fresh as possible when you freeze it. Freezing and frozen storage cannot improve the quality of poor fish; and the shelf life of poor quality fish is not as long as that of good, fresh fish. Poor quality fish also loses more moisture when it is thawed (this feature is called drip loss), which makes the flesh tougher in texture as well as a little less nutritious. The factors affecting freshness were described in Chapter One. There are additional factors affecting quality for freezing.

There are many seasonal variations in the natural quality of fish flesh. During and immediately after spawning, a fish's energy reserves are low, the flesh contains more water than usual and it is relatively soft. Fish frozen in this condition tend to lose more water when thawed and also lose more moisture during storage. Spawning fish are also less flavorful than the same fish when they are in peak condition at other seasons. Salmon are a perfect example of this. Salmon swimming upstream to spawn not only stop feeding, they also expend so much energy swimming against the stream that they burn off most of their fat reserves. What is left is tasteless, watery flesh.

Rigor mortis also affects freezing techniques. Do not fillet or freeze a fish until after it has passed through rigor. If a fish is frozen before rigor, the rigor mortis process will still take place—but very slowly—in frozen storage. If you cook the fish before rigor mortis is complete, the flesh will contract during cooking, lose excessive fluid and toughen.

## Freezing Rate

The water content of fish varies between 60 and 90 percent depending on the species and the season. When fish is frozen, the temperature of the fish's flesh drops to between 28°F and 30°F, which is the freezing temperature of fish muscle. At this temperature, ice crystals begin to form in the tissue. Water expands as it freezes, so these ice crystals take up more volume than water. If the freezing process below this 28-30°F point occurs quickly, then the ice crystals do not have time to grow fully, so they will be small and will not damage the tissues. If freezing is slow, large ice crystals form and can rupture the walls of the tissue cells. A result of this gradual process is that the broken cells, when thawed or cooked, release too much of their moisture and with it some minerals and

> **Box 2: How Freezing Slows Down Fish Spoilage**
>
> Fish spoilage at temperatures above freezing is largely due to bacterial activity. Enzymes naturally present in the fish also contribute. After the fish die, these enzymes slowly digest the flesh, resulting in softened texture and distended or burst bellies. Bacteria multiply faster on the softened flesh produced by the enzymes.
>
> Freezing temperatures (0°F and below) stop the growth of spoilage bacteria and also retard the action of the enzymes in the flesh. This is why freezing is used for preserving food.
>
> Bacteria require "free water" in order to multiply. One reason they do not grow in frozen food is that most of the water is frozen into ice and so is not available for bacterial use. Other preservation methods, such as drying and salting, also preserve by denying bacteria their moisture. In all of these methods, there is too little water present to permit bacteria to function and multiply.
>
> Additionally, freezing destroys 50 to 90 percent of those strains of bacteria responsible for fish spoilage. Be aware that the remaining bacteria, however small the percentage, will start to grow and multiply as soon as the fish is thawed and moisture is available.

soluble proteins, reducing both the flavor and nutritional value of the fish. This also damages the texture of the fish, resulting in a tough, stringy and fibrous flesh. See Box 2 for details on how freezing affects spoilage.

It is vitally important to freeze fish very quickly, especially between 32°F and 23°F, which is the critical freezing zone. If it takes more than two hours for the temperature to pass through this range, the process is described as "slow freezing." There are many different methods of "quick freezing" that can pass the critical temperature zone in less than two hours.

To ensure the fastest possible freezing in a domestic freezer, do not freeze too much at once. It is better to freeze small batches, keeping the rest of the fish well iced in the refrigerator, rather than freezing too much, too slowly. It is also important to keep packages fairly thin and to leave as much area as possible for the cold to penetrate—which means never stack packages that are to be frozen, but always lay them out separately around the freezer.

If you have a freezer with a fan to circulate the air (many no-frost freezers work this way), you can expect better results since the air circulation lowers the temperature of the packages faster.

## Glazing

Glazing is a simple, cheap and very effective way to protect fish in the freezer. Glazing is the term for covering the product in a thick film of ice.

To glaze, first freeze the fish. Whole or dressed fish, steaks and fillets can all be glazed. Once the product is well frozen, dip each piece in a bowl of very cold water, then return it to the freezer to set the coating hard. Repeat this several times so that the glaze is at least $1/16$ inch thick all over.

Hold the pieces in different places each time: you will not glaze where your fingers are, so make sure that these spots get covered next time around.

If you mix one ounce of corn syrup with each quart of water used for the glaze, the coating will be less brittle. But plain water works fine.

Glaze prevents the fish from drying out during storage. It also provides a complete barrier to air and so greatly retards rancidity. After several months in storage, glaze will be much thinner, as the ice evaporates. At that point, glaze again to build the coating back to its original thickness. In this way, the fish will be well protected for the longest possible time.

One way to reduce evaporation and the need to re-glaze is to overwrap the glazed fish with freezer wrap and heavy duty aluminum foil.

## Packaging

Careful packaging is important. It protects the fish by reducing contact with the air during storage, which reduces rancidity and dehydration. Packaging also helps to prevent bacterial contamination from the environment, prevents contamination with odors from other products and makes it possible to label your packages.

Dehydration in storage is obvious from the fish's wrinkled, dry appearance. The affected portions are tough and fibrous when cooked. This condition is known as "freezer burn," and good packaging minimizes the problem. The packaging should be impermeable to water vapor and should cling closely to the surface of the fish (more information later on what type of packaging material to use).

Temperatures inside a freezer fluctuate widely, even though they may remain well below freezing. Opening the door, adding and removing packages and the defrost cycles all cause temperature fluctuations. When the temperature rises, water vapor is forced from the surface of the frozen product. When the temperature falls, this moisture condenses as snow inside the package. Using a skin-tight package helps to minimize these effects.

Rancidity is the oxidation of the natural oils in fish and is caused by contact with the oxygen in the air. Rancid fish tastes objectionable and the color often changes: the fatty strip along the side of the fish may change to a yellow or rusty color. In red-colored fish like salmon and ocean perch, the red skin color may fade to yellowish. To minimize rancidity:

1. Store at as low a temperature as possible.
2. Do not let the fish come into contact with metals, such as iron or copper, which act as catalysts and speed rancidity.
3. Treat the fish with a retardant, especially if it is fatty fish like salmon or tuna, by dipping for 30 seconds in a 3 percent solution of ascorbic acid (vitamin C). Use one measured teaspoon of ascorbic acid to one pint of water or one ounce to three quarts. Synthetic ascorbic acid is the same as natural and many times cheaper. A weak solution of lemon juice can also be used effectively. Lemon juice additionally helps to neutralize the fishy taste that develops in some fish as they get less than perfectly fresh.
4. If you store fish dressed rather than filleted, there is less cut surface exposed to air. This reduces potential oxidation. However, whole fish other than those to be used in pan-ready form take up more room in the freezer, are difficult to package and stack and are far less convenient to use.
5. Freeze your fillets in compact blocks up to about 1½ inches thick. This also reduces the area exposed to the air. If the fillets are interleaved on a strong plastic film, they can be

separated quite easily by banging the sheet on a very hard surface to separate the frozen fillets. Small fish, ready for the pan, can be frozen in a similar way. If you use a carton to hold the blocks, top this up with cold water after freezing the fish, to provide a thick protective glaze.

6. To reduce rancidity in fatty species such as bluefish or mackerel, remove the fatty strip along the lateral line of the fillet, just below the skin. This contains the oiliest part of the fillet, which is the part most likely to go rancid.

7. Use a good oxygen-barrier material to wrap the fish. A good freezer wrap film with an overwrap of heavy foil works well. Standard tinfoil by itself is not recommended for any frozen food.

8. Glaze the fish.

Because most packaging materials provide some insulation, it is best to freeze fish sections unwrapped on trays. This permits the fastest possible freezing. Glaze and package the fish after freezing. Wrap tightly in material that is impervious to water vapor and oxygen. Heavy aluminum foil and freezer-grade kitchen plastic wraps are good.

Wrapping in plastic first, then overwrapping with aluminum foil, provides excellent protection. Make sure that sharp edges or bones do not puncture the wrap, so that the package remains thoroughly sealed.

Brown paper and waxed paper are not satisfactory. Most freezer bags provide good protection against moisture loss, but not against oxygen transmission.

It is important to seal as much air as possible out of the package. If you use bags, immerse the bagged fish slowly, open end up, in a large container of cold water.

The pressure of the water forces out the air at the open end. While the bag is still immersed, tie the open end tightly, remove the package from the water and freeze it.

Watch out for the thickness of packages when you prepare for freezing. Thicker pieces take much longer to freeze and slower freezing means less quality for eating. As the diagram (see Figure 19) shows, if a 1-inch thick piece takes 2½ hours to freeze, a 2-inch thick piece will take about twice as long—nearly five hours. But a 2½ inch thick piece will take three times as long—nearly eight hours.

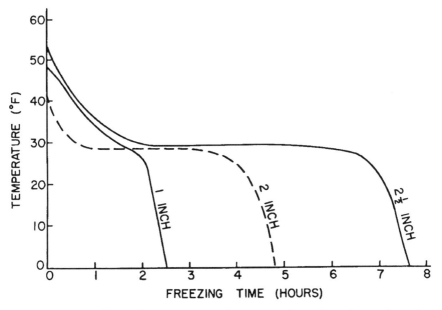

Figure 19. Effect of product thickness on freezing time of packaged fillets frozen in a blast freezer.

## Labeling

Label each package with a description of the product, the date of freezing and an "eat-by" date (determined from the following section on Storage Time). Freezer labels with freezer-proof adhesive are available, although not recommended since many of them fall off in use. Marking the package works much better. Do not use permanent markers since the ink can penetrate the paper and contaminate the fish. Freezer tape, used for strapping up packages wrapped in freezer paper, sticks well, and you can write on it effectively with a felt-tip pen.

## Storage Time

Many books and articles on freezing provide information on how long you can store a product. This chapter adds to the literature. But—and it is a very large but—none of this information is very reliable because of the enormous variations in storage conditions, in temperatures and in temperature fluctuations.

The single most important factor affecting the storage life of frozen fish is storage temperature. The lower the storage temperature, the longer the storage life. Figure 20 shows relative spoilage rates for frozen red hake at a range of storage temperatures. There is a dramatic difference between fillets stored at 20°F and those stored at 0°F. We recommend storing your fish at 0°F or below.

Harder to estimate is the effect of temperature fluctuations at sub-zero levels. These changes, caused by opening doors, defrost cycles, adding additional product to freeze in the freezer, or even power supply problems, are not detectable because the fish stays frozen hard. Fluctuating temperatures produce changes in the size of ice crystals. These changes rupture cells and accelerate dehydration and rancidity development. Stable temperatures are most important for long-term storage. They are also difficult to achieve and to monitor.

Other factors affecting storage life include the condition and quality of the fish at the time it is frozen, the rate of freezing and the type of packaging. Most important,

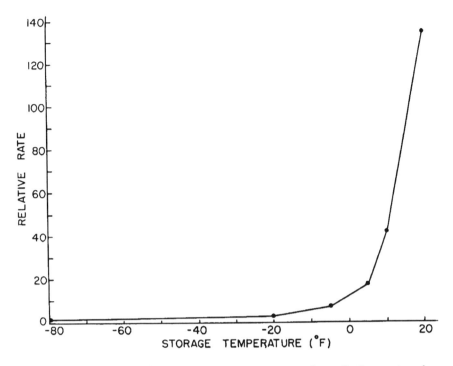

Figure 20. Effect of storage temperature on the relative rate of quality loss in frozen red hake.

different types of fish have inherently different storage lives. In general, fish with a higher fat content, such as bluefish and salmon, do not last as long in the freezer as do leaner fish, such as flounder, whiting and haddock.

The Association of Food and Drug Officials of the USA recommends that you store frozen foods at temperatures below $0°F$. The typical home freezer will usually meet this standard.

Assuming your home freezer operates at $0°F$, Table 3 shows estimated storage life for a number of marine fish and indicates whether flavor or texture will deteriorate first. When you refer to this table, please bear in mind all the warnings mentioned in this section: it is only a general guide and in no way a set of rules. In practice, assume that your home refrigerator/freezer is adequate for no more than half of the times shown in Table 3. This should ensure that you enjoy your catch at its best.

Table 3: Estimated Storage Life at $0°F$ for Various Species of Marine Fish

| Species | Storage life (months) |
| --- | --- |
| Alewives | 5-8 |
| Bluefish | 6-8 |
| Butterfish | 8-10 |
| Cod | 8-12 |
| Dogfish | 4-5 |
| Flounder (sole) | 7-12 |
| Haddock | 9-12 |
| Halibut | 7-10 |
| Herring (sea) | 3-5 |
| Mackerel (Atlantic) | 4-6 |
| Ocean perch | 7-9 |
| Pollock | 8-10 |
| Red hake | 6 |
| Salmon | 5-9 |
| Scup | 8-12 |
| Smelt | 5-9 |
| Spanish mackerel | 4-6 |
| Striped bass | 8-9 |
| Squid | 12-24 |
| Tuna (bluefin) | 4-6 |
| Weakfish | 6-8 |
| Whiting (silver hake) | 7-10 |

Source: National Marine Fisheries Service.

## Thawing

The best way to use frozen fish is to cook it from frozen. You can use fillets, steaks and small whole fish this way, and the results are better than if you thaw the fish first. When preparing fish for the freezer, remember that if you prepare portions suitable for your kitchen (such as steaks, fillets or even smaller portions), you can avoid the need to thaw them before you cook and eat them.

If you must thaw, say for a recipe calling for rolled fillets, microwaving is a simple and effective method. Follow the instructions in the microwave handbook. Every model is different and there are no universal rules. Microwaves are particularly useful if you have a block of fillets or small pan fish and need to defrost the block sufficiently to separate the fish so that you can use some and put the rest back in the freezer.

The next best method of thawing is to leave the fish overnight in the refrigerator so that it thaws as slowly as possible. Quicker thawing at room temperature or higher leads to greater drip loss. Also, thawing too quickly produces fairly high temperatures on the surface where bacteria can grow more readily. Avoid thawing in running water. Although this method is fast, it also washes away moisture, flavor and nutrients.

## Refreezing

You can refreeze thawed fish, provided that it was thawed for only a short time while kept at refrigerated temperatures. Watch for two things:
1. Refrozen fish often have poorer texture and flavor than once-frozen, because of the additional strain on the cell structure of the muscle

and additional drip loss. This applies to most fish, although you can refreeze squid, for example, several times without a noticeable loss of quality.
2. Bacteria do grow on the thawing fish. It is important not to refreeze a product if you are uncertain whether it was held in clean and cold conditions while defrosting.

## Summary

Freezing fish is one of the best ways of preserving part of your catch for later use.
1. Use good, fresh fish. The quality cannot improve in the freezer.
2. Clean and prepare the fish so that it is ready to season and cook when you take it out of the freezer later.
3. Package the fish so that it is as airtight as possible. Glazing helps in preventing dehydration and delaying rancidity.
4. Take care not to freeze too much at once. An overloaded freezer will take too long to freeze the fish. Fast freezing is essential.
5. Cook fillets, steaks, portions and pan fish from the frozen state. It is not usually necessary to defrost them first.

*Chapter Three*

# Salting Fish

Salt was a precious commodity over 2000 years ago when it was taxed in China. The Romans valued it highly; the word salary comes from the Latin *salarium*, the salt money given to soldiers to enable them to buy this essential food preservative.

Salting of fish is also ancient. In 3000 B.C. the Egyptians were exporting salt fish, long before the European market for salt cod and herring impelled the settlement of parts of New England and the Canadian Maritimes.

The ancient Egyptians had very pure salt. The early American settlers were less fortunate and had to import salt from Spain and the Caribbean. There are 18th century records complaining about the poor quality of New England salt fish—"salt burnt" because of impurities in the salt used.

For salting, pickling and smoking, it is absolutely essential to use only the purest salt. Calcium or magnesium salts and almost all sulfate, while not harmful, delay the penetration of the salt into the flesh. This delay allows bacteria to grow, and the fish begins decomposing.

Salted cod and herring, once important staples of world trade and major causes of wars, are now expensive gourmet foods. It is no longer necessary for survival to have salt cod or a barrel of herring in the pantry. But it is definitely a treat to have access to these products. There are numerous recipes for using salt fish, and though many recipes are for salt cod, you can substitute other lean fish such as pollock, haddock, sea bass, catfishes, rockfishes, saugers, walleyes, yellowtail or cusk. Similarly, you can adapt recipes using salted or pickled herrings to other fat fish like mackerel, albacore, bluefish, salmon, tunas, alewives and shad.

Salt fish keeps almost indefinitely in cool, dry conditions. Its one disadvantage is that it is not as easy to cook as frozen fish, because you have to soak it first to remove the salt. After you remove the salt, the possibilities for creating great meals are endless.

## How Salt Preserves

Salt replaces the moisture it removes from the flesh. Bacteria cannot grow without plentiful moisture and the salt also inhibits their growth. Once the salt content of the flesh tissues rises above 9 percent, the effects of most enzymes and bacteria halt. A salt content of 20 percent preserves fish for a very long time, unless it is stored in hot, wet conditions.

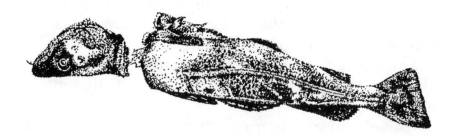

Figure 21.   Scale and behead fish.

## Preparing the Fish

1. Carefully bleed and gut your catch (see Chapter One). If you are salting your fish, bleeding is particularly important to ensure a pleasingly light colored product.
2. Scale and behead the fish (see Figure 21).
3. Cut the belly side from the vent along the left side of the backbone down to the tail, but do not go any deeper than necessary for cutting out the backbone. Traditional salt fish is based on split, headless raw material (see Figure 22).
4. Cut directly across the backbone about three-fifths down from the head end. Do not break the backbone at this point, but make sure it is cut.
5. Hold the backbone with one hand and cut from the point of this cut under the backbone away from the head, keeping the blade of the knife as close as possible to the backbone. If you do otherwise you will waste too much flesh, which will cling to the backbone (see Figure 23).
6. Remove the forward part of the backbone, leaving in about two-fifths in the tail section (see Figure 23).
7. Remove the membrane covering the inside of the gut cavity.

When finished, the split fish should resemble an elongated triangle (see Figure 24).

An easier preparation method, most suitable for fish over 10 pounds or so, is simply to fillet the fish and leave the skin on.

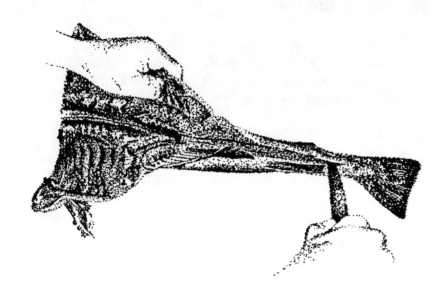

**Figure 22.** Cut belly side from the vent along the left side of the backbone down to the tail.

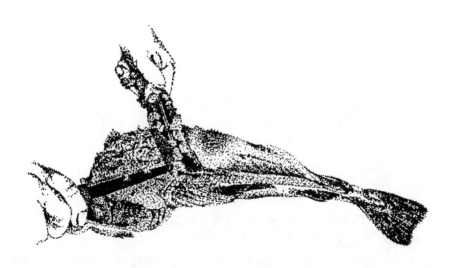

**Figure 23.** Cut from the backbone away from the head.

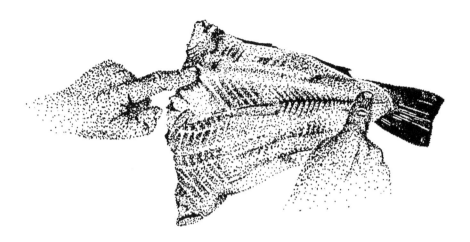

Figure 24. Split fish should resemble elongated triangle.

Whatever method you use, before salting, inspect the fish carefully and remove any blood spots. Rinse well to eliminate any slime, loose scales or loose bits of flesh. Do not soak the fish as this will soften it.

## Salt

It is essential to use pure salt, without metallic or other salts. Sulfates, magnesium salts and calcium salts slow the penetration of sodium chloride into the flesh and also leave a bitter taste. Sea salt is not suitable unless it is particularly pure. For small quantities you may use ordinary table salt that is not iodized. Kosher salt, readily obtainable in supermarkets, is pure and highly effective. Purchase 100-pound bags of meat curing or canners salt if you have large amounts of fish to cure. These are much cheaper than ordinary household salt. You can find suppliers in the yellow pages. Avoid rock salt.

## Pickle Curing

A nonmetallic container is a must for pickle curing. Use a ceramic crock, plastic, glass or even a watertight wooden barrel. Avoid all metals (except stainless steel) because they react with the salt and contaminate the fish (see Chapter Four for more information on pickling and marinating).

First put a generous layer of salt over the bottom of the container. Then add a layer of fish, flesh side up. Do not overlap the fish: it is important to have the entire surface covered with salt. Spread salt evenly over the fish, leaving no uncovered spots. Then add another layer of fish and continue until you use all the fish. Stack the top three layers of fish skin side up and coat the top of the stack with a thick layer of salt.

Put a wooden board or a plate on the top and add something heavy enough to keep this under the surface of the brine as it develops. The objective is to keep the fish below the level of the solution at all times. Plastic containers filled with water make suitable weights as do clean jars filled with water (substitute a nonmetallic lid when using jars).

The salt extracts water from the fish to form brine. This brine should test between $90°$ and $97°$ on a salinometer. A salinometer is a hydrometer designed to measure the specific gravity of brine. Pharmacies should be able to supply one at modest cost. If the reading is above $97°$, the fish will have poorer texture and flavor. If the reading is below $90°$, the fish may spoil. A traditional method for measuring the approximate brine strength is to float a raw, peeled potato in the brine. If it sinks, add more salt. If it barely floats, the brine strength is considered correct. However, this test is not too reliable. Use the brine table to determine how much salt you need to add to correct the brine mixture.

Table 4: Brine Strengths Measured at 60°F and 65°F

| Salinometer Degrees | % Salt by Weight | lb. Salt/Gallon of Water 60°F | lb. Salt/Gallon of Water 65°F |
|---|---|---|---|
| 0 | 0.000 | 0.0000 | 0.0000 |
| 10 | 2.640 | 0.2168 | 0.2260 |
| 20 | 5.279 | 0.4456 | 0.4640 |
| 30 | 7.919 | 0.6880 | 0.7160 |
| 40 | 10.558 | 0.9440 | 0.9830 |
| 50 | 13.198 | 1.2160 | 1.2660 |
| 55 | 14.517 | 1.3592 | 1.4150 |
| 60 | 15.837 | 1.5064 | 1.5680 |
| 65 | 17.157 | 1.6568 | 1.7250 |
| 70 | 18.477 | 1.8136 | 1.8880 |
| 75 | 19.796 | 1.9744 | 2.0550 |
| 80 | 21.116 | 2.1416 | 2.2290 |
| 85 | 22.436 | 2.3144 | 2.4090 |
| 90 | 23.755 | 2.4920 | 2.5940 |
| 95 | 25.075 | 2.6768 | 2.7870 |
| 100 | 26.395 | 2.8696 | 2.9870 |

## How to Salt Fish

For this operation, you need about 1¾ pounds of salt for ten pounds of fish. Small fish or thin fillets require a little less salt than thicker ones. Use a little more salt in warmer weather if necessary. Cod, hake, cusk, pollock and other bottom fish are suitable for this method. Lean fish (less than 1 percent fat) are best for salting because their flesh holds up well without breaking or flaking into pieces.

You can also salt fatty fish such as mackerel, salmon, herring and trout, but there are special requirements. See the section later on in this chapter on "Salting Fatty Fish."

The fish will be ready to use in three weeks. Provided you keep the fish below the surface of the brine and keep them reasonably cool, they will hold well in the brine for as long as a year.

## Soaking Salted Fish

Use brine-cured fish straight from the brine, without further processing. Soak the fish thoroughly in several changes of cold, fresh water before using. Soak for up to 24 hours, preferably in a refrigerator, using at least three changes of water and rinsing the fish between each change. This removes most of the salt. Without soaking, any dish made from the product is unacceptably salty. You can use all recipes for salt fish or salt cod for this product.

Traditionally, salt fish is dried after being pickle cured as described above. Drying at home is seldom practicable and is not really necessary since you can use the fish in exactly the same way, whether it is dried or not. The following instructions are for those interested in producing dried fish.

## How to Dry Salted Fish

Once the fish is brined (after at least three weeks), remove it and rinse with a weak brine solution (about $20°$ to $30°$ on a salinometer) to remove excess surface salt and any slime or dirt that may have accumulated.

The next part of the process is called "kenching," which means piling or stacking. The purpose is to press out some of the moisture of the fish and to give a smooth surface to the fish. You need a slatted platform or rack, with some way of draining the fishy brine that seeps underneath. Place two layers of fish, skin side down, on the rack. Then pile on the remaining fish flesh side up. If the pile is not very great, add some weights to help squeeze out the moisture. After one day, reverse the order in which the fish are stacked, so that the top of the pile becomes the bottom. Generally, the process is complete within two days. At this point the fish are called "green-salted."

## Requirements and Conditions for Drying

The final phase is drying, which is easy to describe but much more difficult in practice because of the problem of finding a suitable location. You can dry your fish outdoors, on a rooftop or in the backyard. You can also dry your fish indoors, in a dry cellar or attic. Either way, you need a cool, dry place with plenty of airflow. If you do it indoors, you should be prepared for plenty of less than pleasant odors. If you do it outdoors, you must protect the fish from raccoons, dogs, cats, birds, bird droppings, insects, rainstorms and irate neighbors.

The aim of drying is to reduce the water content to about 50 percent. Indoors, a low volume of air from a fan running at low speed is ideal for drying. Do not use a high fan speed as this creates a crust on the outside of the fish and the inside does not dry properly. Temperatures between 65°F and 75°F are ideal. At least 10 hours are necessary to dry the fish under these ideal conditions. Outdoor drying simply involves exposing the green-salted fish to light breezes and some (but not too much) sun at similar air temperatures.

Expose the fish to air on both sides. Attaching it to chicken wire, or even a clothes line, works fine. Box 3 gives definitions and descriptions of the many different types of salted and cured fish.

## Storing Air–Dried Fish or Fillets

Although this product can keep as it is (you can even store it in a drawer, provided the humidity is not too high and insects and rodents are kept away), it is generally preferable to skin and debone the fish so that it is ready to use. Package it in portion-size packs of plastic wrap or in plastic bags. A cool, dark place is best for storage. The kitchen refrigerator is ideal.

## Salting Fatty Fish

Air and sunlight turn fats rancid. Therefore, you must brine fatty fish in opaque containers so that light cannot reach the contents. Do not dry fatty fish: instead, use it directly from the brine.

Mackerel are an excellent salted fish, readily available to marine anglers in both the Atlantic and the Pacific. So is salmon, now abundant in the Great Lakes as well as on both coasts. Use only the freshest fish (and those that were well-iced when landed) for salting. To ensure top quality product, fillet (see Chapter One) and trim the fish, then wash them to remove all traces of blood and any particles of loose flesh before salting.

Rub the fillets gently in salt so that they pick up as much salt as possible. Then put them in a container, just as described above for lean fish. Pack the bottom layer with the skin side down, the next layer with the skin side up. Alternate layers to keep the fish as level as possible. If you are using a circular container, pack the fillets in a circular fashion with the tails towards the center. Scatter a thin covering of salt on each layer of fish so that the entire surface is in contact with the salt. Finish with a

## Box 3: Types of Cures

**Amarelo:** Portuguese cure. Salt content about 18 percent. Yellow color.

**Branco:** Portuguese cure. Salt content about 20 percent. White color.

**Dry salting:** See Kench, below.

**Fall:** Light salt, pickle cure, 45 percent to 48 percent moisture.

**Gaspé:** Light salt, pickle cure, 34 percent to 36 percent moisture.

**Hard:** Dry salted; dried to moisture content of 40 percent or less.

**Heavy salted fish:** Has about 40 percent salt (dry weight basis) and moisture content with a range of definitions, in Canada, as follows:
  **Extra hard dried** — less than 35 percent moisture
  **Hard dried** — up to 40 percent moisture
  **Dry** — 40 to 42 percent moisture
  **Semi-dry** — 42 to 50 percent moisture
  **Ordinary cure** — 44 to 50 percent moisture
  **Soft dried** — 50 to 54 percent moisture

Heavy salted soft cure fish have about 17 percent salt (dry weight basis) and about 47 percent moisture.

**Kench:** Dry salting: Split fish and salt in alternate layers allowing the moisture to drain. This is a basic curing method.

**Labrador:** Heavy salt (about 18 percent); 42 percent to 50 percent moisture.

**Pickle cure:** Wet salting: the fish are salted in a container, so that they are cured or pickled in the liquid that forms.

**Shore:** Light salt, kench cure. About 12 percent salt and 32 percent to 36 percent moisture.

**Soft:** Dry salted, dried to over 40 percent moisture.

The following are some **examples of dried fish:**
  **Stockfish:** Unsalted fish dried in cold air. Fish is headed, gutted and hung to dry in the sun.
  **Dried salted codfish** (also called bacalao and klipfish): Salted fish is washed, salted again and dried (see Saltfish, below). Keeps well in cool, dry areas.
  **Saltfish:** Headed, gutted with the larger part of the backbone removed. The fish is cleaned and washed, then preserved with dry salt for three weeks. Sold whole or in fillets, it keeps well in cool temperatures.

Source: *The New Frozen Seafood Handbook* by Ian Dore, Van Nostrand Reinhold/Osprey Books, 1989.

layer of salt. Put a weight on the surface, as you did with the lean fish, so that the fish remain under the brine as it forms.

Early in the spring, when both mackerel and coho salmon are relatively lean, use about one pound of salt for four pounds of fillets. When the fish are high in fat later in the season, use one pound of salt to about three pounds of fillets.

Mackerel and similar size fish should be brined in about 10 days if they are fairly lean and in up to three weeks if they are at their fattest. Make sure the container is not too large for the quantity of fish: the brine MUST cover the fish at all times.

Large fatty fish such as tuna, bonito and chinook salmon can be handled similarly except that the fish must be cut into small pieces. Fillet first, then cut the fillets into strips lengthwise, then cut crosswise so that the final pieces are about 6 to 9 inches long and about 2 inches wide—roughly the size of mackerel fillets.

You can keep fatty fish in the brine for at least six months, provided you maintain the brine strength between $90°$ and $97°$ and the temperature at no more than $60°F$. Temperatures rising above $60°F$ reduce safe storage to about three months. Make sure to keep the fish in the dark.

## Some Helpful Hints for Salting Fish

- Never use metal weights for pressing the fish into the brine. Use clean, heavy stones instead.

- Do not let iron, aluminum or copper come into contact with the fish or the brine.

- Jelly or pie-filler 5-gallon plastic pails from bakeries make good containers for salting fish.

- Do not use transparent plastic containers as light can adversely affect the fish being cured.
- Fish is cured when the thickest part of the piece feels hard and is highly resistant to finger pressure.
- "Freshen" the fish before use by soaking for about 12 hours (sometimes up to 24 hours) in tap water. Change the water at least three times and rinse the fish well each time. Keep the fish in the refrigerator while it is soaking. Then use the freshened, salted fish in any salt fish recipe.

## Recipe for Gravlax

This is an easy way to preserve and enhance fresh salmon. It tastes similar to the very best smoked salmon, but is more delicate and far easier to produce.

Atlantic salmon, steelhead and Pacific king (chinook) salmon make the best gravlax, but silver, chum, sockeye and pink salmon are all entirely acceptable alternatives.

### *Ingredients*

    4 to 5 pounds of fresh salmon
    A large bunch of fresh dill
    ¼ cup of kosher salt
    ¼ cup of sugar
    1 tablespoon of crushed white peppercorns

Fillet the fish. You need both fillets, or (even better) the center cuts from both fillets. Remove the pin bones from the center line of each fillet with pointed-nose pliers, pulling each bone out with a sharp tug.

Mix the salt, sugar and crushed pepper together in a small bowl. Put a thin layer of the salt/sugar mixture over the bottom of a glass, porcelain or plastic dish (do not use metal). Lay one fillet, skin side down, on the seasonings. Sprinkle more salt/sugar mix over the flesh side of the fillet. Cover this with a layer of torn pieces of the fresh dill weed. Put the second fillet over the dill, skin side up, with the thick part over the thin end of the first fillet—that is, put the fillets so that they are nose to tail, flesh to flesh, with the thick part of one against the thin part of the other. Sprinkle the rest of the salt/sugar mixture over the top. Cover the dish with foil and put a weight on the fish. Another dish, plate, or a board with several full cans on top works well. Refrigerate for at least two days, turning the fish every 12 hours. Each time you turn the fish, baste it with the marinade that develops in the dish.

The gravlax will be ready in two to three days and will keep for about a week if the salmon was really fresh. It can be frozen and keeps extremely well in the freezer. Make sure it is tightly wrapped.

To serve, slice thinly on the diagonal from the top down towards the skin. Slice the fish off the skin. Gravlax is a delicious alternative to smoked salmon, either as hors d'oeuvres or as a main course.

The same recipe works well on mackerel. Butterfly the fish, removing the backbone, then proceed as for salmon, laying each fish alternately head to tail.

Chapter Four

# Pickling and Marinating Fish

Preserving and enhancing fish by using vinegar or other acids is an ancient, effective and particularly palatable technique. In this chapter, we offer ideas for using vinegar and marinades for preserving fish. The techniques we suggest all produce good, appetizing dishes.

Vinegar ("sour wine") has been made for at least 10,000 years. Pickling with vinegar was widely practiced by the Greeks and the Romans. Pickled fish products are important in many European countries. In eastern Asia, especially Japan, coastal China and the Philippines, pickling is an extremely important method of fish preservation.

Vinegar preserves fish because it makes the flesh too acidic for spoilage bacteria, which flourish in a low-acid environment. It also coagulates the protein, just as heat does when you cook. When salt is used with vinegar, you get preservative effects from both. Spices, normally used with vinegar in pickled and marinated products, have no preservative effect but enrich the flavor.

Pickling and marinating are extremely flexible techniques, because you can almost infinitely vary the spices, herbs and other flavorings to suit individual tastes.

You can satisfactorily pickle or marinate most fish. Herring are the fish most often used, but since they do not take hooks, anglers do not normally catch them. If you use herring for bait, consider keeping some of them as food for yourself and your family (but only if the herring are really fresh.) Herring are an excellent, tasty and very nutritious fish. You can substitute alewives, shad and other fish closely related to herring in most herring recipes. Other fatty fish are equally good. Mackerel, which anglers can readily catch, make great marinated and pickled products. Other usable fish include salmon, tuna, chub (Pacific), king and Spanish mackerels and bluefish. You can also pickle lean fish like bullheads, cod, ling, walleye, haddock and pollock.

The most important feature of the fish you pickle is quality. The flesh must be fresh, clean and free of all traces of blood, slime and viscera. Poor quality fish result in unsatisfactory, generally mushy, pickled products.

## Ingredients for Pickling and Marinating

**Vinegar** for pickling should always be clear, white, distilled vinegar. Avoid cider, wine and tarragon vinegars for the first preservative stage of pickling. You may use flavored vinegars at a later stage for making other products, but if you use them for the basic pickling, an unpleasant flavor develops. Check the label to determine that it says "distilled vinegar" and that it states that the strength has been reduced to 5 percent acidity or to 50 grain strength (which is the same thing).

Make sure **spices** are fresh and in sealed containers. Opened spices lose their potency quickly. Invest in the small cost of new spices to ensure the best flavor.

The **salt** you use must be pure salt without metallic or other salts. It cannot be iodized or contain sulfates, magnesium salts or calcium salts. These minerals delay the penetration of the salt into the flesh. Kosher salt is the easiest pure salt to obtain. Do not use salt substitutes or potassium chloride. Chapter Three contains more detailed information on salt and on salting fish.

**Sugar,** when used, is ordinary household white (cane) sugar.

**Onions,** when specified, are ordinary yellow onions. The more expensive Bermuda and Spanish onions are bland and less satisfactory.

**Water quality** is most important. Hard water is often high in calcium, iron or magnesium. These minerals may discolor the fish and spoil the flavor. If you live in a hard water area, filter or treat the water, use a water softener or buy bottled water without minerals to use for pickling.

Everyday household **utensils** are generally usable for pickling. There are, however, a few materials you should avoid. Aluminum pans darken and sometimes pit when used with vinegar. Do not use cast-iron pots unless they are enameled and there are no chips in the enamel. Heat-proof glass or ceramic pans are the best for pickling.

## Recipes

Many of these recipes are traditional European ones that have been successfully used for many years. You can vary proportions to suit larger or smaller quantities.

### Uncooked Marinade

A favored appetizer is an uncooked marinade that uses herring as a base and wine as a flavor enhancer. Anglers can use other species from their catch such as mackerel, trout, salmon, catfish or any firm-fleshed white fish.

Fillet and skin enough fish to yield ten pounds of fillets. Cut the fillets into pieces about two inches square and not more than two ounces in weight. Wash and drain well. Place the fish in a $95°$ salinometer brine for about four hours (see Chapter Three for details of how to make and use brine and follow those instructions on keeping the fish under the surface of the brine). Remove, drain and rinse the fish.

Mix four cups of distilled white vinegar with 1½ cups of water. Cover the fish with this mixture (make more to the same proportions if required), holding the fish under the surface with a suitable weight such as a clean glass jar filled with water. Leave the fish in the vinegar solution for 48 hours. Remove the fish, pack it into jars and cover with a wine-based sauce such as the following:

### Wine Sauce for Uncooked Marinade

- 1 quart dry white wine
- 1 pint distilled white vinegar
- 1½ cups chopped onions
- ⅓ cup sugar
- 2 tablespoons crushed red peppers
- 2 tablespoons whole cloves
- 1¼ tablespoons whole black peppers
- 1¼ tablespoons whole white peppers
- 2 tablespoons whole allspice
- 1½ tablespoons ground nutmeg
- 1 stick cinnamon
- 1 tablespoon crushed cardamon pods (optional)
- 15 whole bay leaves
- 1½ tablespoons mustard seed

Place all the ingredients in a double boiler. Wrap the spices in a cheesecloth and keep them suspended within the liquid so that they do not burn at the bottom of the boiler. Boil for two hours. Let the mixture stand overnight so the flavors blend completely and the marinade cools. Use the wine sauce to fill the jars of fish. Keep the jars in a cool, dark place.

The product will keep six months under refrigeration and about three months in a cool (not refrigerated) dark area such as a cellar. Temperature should be 60°F or less.

Vary the spices to suit personal tastes. Experiment with different spices, different proportions and red wine instead of white. The recipe given here is only a guide. Each combination of spices gives you a different product. Find and stay with ones you like.

## Seviche

Seviche is made of bite size pieces of fish marinated in lemon or lime juice. Traditional seviche consists of corvina or bay scallops, but you can adapt this recipe by using almost any type of fish, although lean fish generally result in a better product. Some fish that work well in this recipe are black bass, walleye, red snapper, weakfish and lake trout. Squid is also excellent.

Begin with skinless fillets sliced crosswise into ½ inch strips, then cut them into ½ inch cubes. Place the cubes in a glass or stone container, separating each layer of fish with onion rings, flaked red pepper, black peppercorns and a bay leaf. The amount of spices you use is entirely up to you. You can experiment with additional spices and seasonings for a different taste.

Marinate the mixture in lemon or lime juice. An alternative to these juices is a half-and-half mixture of white vinegar and citrus juice. Refrigerate for 24 hours. The acid "cooks" the fish, changing the color of the fish from translucent to white.

## Marinated Haddock

Fillet and skin haddock, or any fairly large, lean fish, and cut the fillets crosswise into pieces roughly two inches square. Wash, drain, place in a suitable container and cover with a 95° salinometer brine (see Chapter Three,

Salting Fish). Let stand three to four hours. Drain and rinse. Use a nonreactive (ceramic or enameled or ovenglass) cooking utensil.

*Ingredients*

    10  pounds haddock pieces
     2  quarts distilled vinegar
     1  quart water
     3  cups sliced onions (thin)
     5  tablespoons (½ ounce) white peppercorns
     2  tablespoons (¼ ounce) whole cloves
     2  tablespoons (¼ ounce) mustard seed
     5  tablespoons (½ ounce) whole allspice
     2  tablespoons (½ ounce) crushed red chili peppers
  About 15 bay leaves
  Lemon and onion slice for garnish
  Additional spices

Layer ingredients as follows: first, cover the bottom with a scattering of sliced onions and spices. Next, add a layer of fish. Scatter more slices of onions and spices, then more fish, until you use up all the ingredients. Cover with a mixture of two parts of vinegar and one part of water. Bring to a boil slowly and simmer until you can pierce the fish easily with a fork.

Remove the fish gently, cool, drain and pack in glass containers. Garnish each glass jar with a few fresh spices, a slice of lemon, a slice of onion and a bay leaf. Strain the vinegar cooking mixture, heat to boiling point and pour hot into the jars to cover the fish. Seal immediately. If refrigerated, it will keep for three months. Otherwise, store in a cool, dark place and use by the end of the sixth week.

You can use the same recipe with other kinds of lean, firm-fleshed fish such as catfish, croaker, rockfish, sauger, walleye, yellowtail and many others.

## Marinated Mackerel

- 10 pounds mackerel pieces
- 2 quarts distilled vinegar
- 3 pints water
- 2 cups chopped onions
- 4 tablespoons sugar
- 2 cloves garlic, chopped
- 1 tablespoon whole allspice
- 1 tablespoon whole black peppers
- 1 tablespoon whole cloves
- 1 tablespoon ground nutmeg
- 6 bay leaves

Fillet fresh mackerel and thoroughly rinse the fillets in fresh water to remove any blood. Cut fillets crosswise into two-inch pieces and dredge them thoroughly in kosher salt so they pick up as much salt as possible. Place in a nonmetallic container and let stand for two hours. Then rinse with fresh water. Combine vinegar, water and other ingredients, bring to a boil, then simmer for 10 minutes. Add the fish and cook slowly for another 10 minutes after the liquid comes back to a boil. Do not overcook. Remove the fish and let it drain. Then pack in sterilized jars. Garnish jars as desired with chopped onions, a few spices and a bay leaf or two. Strain the spiced vinegar and bring to a boil. Fill jars with the boiling sauce and seal promptly. Store in a cool place.

## Pickled Fish

- 25 pounds fat fish such as herring, mackerel or bluefish
- 2½ pounds table salt
- 6 quarts 5 percent white distilled vinegar
- 3½ ounces whole pickling spices
- 5 pounds sugar
- 6 pounds yellow onions

Clean fish thoroughly. Remove scales, heads, fins, tails and viscera; be especially careful to remove all of the kidney, which lies beneath a membrane along the backbone. Wash thoroughly and cut crosswise into 1½ inch steaks.

Make a saturated brine solution by adding salt to water until no more salt will dissolve and the excess settles to the bottom (about 2½ pounds of salt per gallon of water). See also Chapter Three, Salting Fish. Pack the cut fish in the container with brine so that the fish are completely covered with brine (and weight them down so they stay under the brine). Keep the fish in the brine container in a cool room, preferably refrigerated to about 40°F. The higher the temperature, the quicker the salt penetrates into the fish. Too high a temperature causes the fish to soften.

Prepare the pickle in advance while the fish sits in the brine. Add 3½ ounces of whole pickling spices and 5 pounds of sugar to the 6 quarts of 5 percent distilled white vinegar. Then heat to about 160°F (hot, but not boiling). Let the pickling mixture cool or refrigerate it.

Meanwhile, leave the fish in the brine until the salt "strikes." You can recognize fully salted (struck) fish by a change in the flesh color from a reddish to gray or tan color completely through each piece of fish. Stir the fish every day or two and add an additional cup of salt every other day to maintain the proper salt concentration. High concentrations tend to wrinkle the skin.

Before using the brined fish, soak the fish in fresh water for two days to freshen it, making sure to change the water several times each day.

Drain the fish from the fresh water and cover it with the cooled pickling mixture. Cover the container and set aside in a cool area or refrigerator, stirring daily. After three days, add five to six pounds of sliced onions. Continue to stir daily. Mackerel or herring pickled in this manner attain their best flavor after three to four weeks of cold storage. Taste after two weeks and add some sugar (or not) according to preference.

Chapter Five

# Smoking Fish

You should never consider smoking as a way of preserving your catch. Although prolonged smoking combined with heavy salting was originally an important preservation method, today's tastes favor lightly smoked and lightly salted products. Processing to meet these tastes does not add storage life. It does add flavor and improves texture and makes fish into gourmet delicacies. You can freeze fish after you smoke it to preserve it, but do not expect it to keep simply because it is smoked. Regard smoking as a way of enhancing the appeal and value of your catch. Preserve the product—raw or smoked—by freezing it.

You can smoke just about any freshwater and saltwater fish and shellfish. There are only a few species that you cannot smoke, one being whiting (also known as Pacific hake) from the West Coast. This species contains an enzyme that breaks down the flesh during the smoking process. (See Chapter Six for more information on whiting.) Anglers have the advantage of access to the best quality fresh fish. Good raw material is essential for producing top quality smoked fish.

## Equipment for Smoking Fish

The equipment you use for smoking is simple. If you are in a remote area, salt the fish in plastic bags, hang them over sticks to dry and smoke them in a lean-to covered with a cloth tarpaulin or some green boughs. Alternatively, you may use a similarly covered tripod of poles over a shallow pit containing smoldering green hardwood. You can also use portable smokers that are widely available at sporting goods stores.

Another alternative is a large, covered backyard barbecue for hot smoking, provided you can control the ventilation to keep small amounts of sawdust or wood chips smoldering. Put the fish on the side opposite the fire, away from direct heat.

If you are more ambitious, you can find numerous plans for constructing backyard smokers from barrels, oil drums, sheds or old refrigerators. *McClane's New Standard Fishing Encyclopedia* has a section on smokehouse cookery with detailed instructions on building smokers. For cold smoking, build the fire six feet from the smoker and duct the smoke through a covered trench in the ground.

The best types of wood for smoking are nonresinous hardwoods such as oak, hickory, beech, alder, sweet bay and buttonwood. Wood from fruit trees gives good color, scent and flavor to fish. Dry corncobs are sometimes recommended, but watch these carefully because they tend to flare up and throw off too much heat. Avoid soft, resinous woods such as pine. These deposit bitter, tarry flavors in the fish.

Sawdust is the best form of fuel for backyard smoking because it is controllable. The burning rate of wood chips, sticks or split wood is more difficult to control, so these materials need a lot of attention to prevent flare-ups and premature dying. Pile small amounts of sawdust on a steel sheet or in an iron frying pan and set to smoldering with an electric hot plate, a propane torch or a piece of burning charcoal.

As the sawdust burns down, brush aside the ashes and add fresh sawdust to the smoldering fire. Experience soon teaches the correct amount of sawdust to use.

Use thermometers, preferably dial types with metal stems, to monitor temperatures and help ensure good, even smoking. Avoid glass thermometers as they are easily broken, especially when they are hot. Mount one thermometer, which should read up to 250°F, through the wall or door in the upper half of the smokehouse to monitor the temperature of the smoke-air mixture. Use a second thermometer to check the temperature of the fish during hot smoking. Insert the probe into the center of the thickest part of the fish.

For drying and smoking fillets and steaks, make shelves covered in ¼ inch to 1-inch hardware cloth with a wooden frame. Coat these screens with edible oil before each use. Move the fish during processing to minimize the marks of the mesh. Hang dressed fish on wooden dowels or steel rods, using S-shaped hooks. Tie the tails of small fish together in pairs and drape the fish over a rod. Thread sharpened dowels or rods through the gills, or through one gill and out the mouth. Using toothpicks or small pieces of wood, prop open the belly cavities of the fish hung for smoking.

Headed and split fish and fish sides can be hung on S-hooks or string inserted beneath the bony neckplate. Another alternative is to drive nails through a 2-inch-square stick and impale the fish on the nails.

## Preparation

If you intend to dry and smoke your catch, it is important to bleed the fish as thoroughly as possible as soon as it is caught. Blood spots on smoked product, although harmless, are not appetizing. Salmon is particularly susceptible to showing dark spots and patches. Fish can be smoked in almost any form. Dressed, headed and dressed, fillets and chunks are all appropriate. Fillets present the greatest difficulty since they need to be laid down in the smoker. If you leave the nape bones on when you behead the fish, you retain a useful handle for hanging sides or fillets. Large fish such as swordfish, shark, mahi-mahi, yellowtail and albacore should be chunked into pieces about 2 inches thick.

After cutting the fish, wash it in a weak brine solution (about one ounce of salt for each pint of water). Soaking in a similar solution for about 30 minutes helps to wash out blood.

## The Smoking Process

The smoking process consists of three distinct operations: salting, drying and smoking.

### Salting

Chapter Three discusses salting in detail. Read that first to understand the basic principles and techniques. Use only noniodized salt: kosher salt or food-grade salt as recommended for salting. You should never use iodized salt for salting fish. However, when salting before smoking, you can use iodized salt if nothing else is readily available. It is better to use plain salt for salting fish under all circumstances.

Fish are salted to enhance the flavor of the finished product and to draw some water from the flesh. After salting, the fish contain three to four percent salt and show a small weight loss. As they dry and lose moisture during the remainder of the process, the salt concentration automatically increases. Note that salt concentrations as low as five percent retard bacterial decomposition, but are not high enough to prevent molds from growing. Once again, it is important to bear in mind that modern smoking is not a preservative, but a flavor-enhancer.

*Sodium Nitrite.* Salt mixes formulated specifically for smoking may contain other ingredients such as brown sugar, sodium nitrite and condiments. Sodium nitrite is used commercially as a preservative to prevent botulism, which is caused by a deadly toxin produced by botulinus bacteria. However, nitrites are also very poisonous, and you must handle and use them with great care and precision. We recommend very strongly that you avoid using nitrites altogether.

Sodium nitrite is used because it protects the fish during distribution when the manufacturer is not able to control what happens to his product. He does not know whether it will be kept properly refrigerated at all times by distributors, retailers and consumers. Producing smoked fish yourself gives you a substantial advantage: you have control over what happens to the product at all times until you eat it.

The most satisfactory answer for the angler smoking his catch is to treat the smoked fish exactly as if it were fresh fish. That is, keep it well refrigerated and assume that the shelf life of the smoked product is exactly the same as if the fish were not smoked. Then preserve smoked fish by freezing. A good product will remain a good product. Additives are not necessary. Note that freezing tends to highlight saltiness, so salt for a shorter time in the initial part of the process if you plan to freeze.

If you should decide to use sodium nitrite, which is available from local chemical suppliers, 1.33 ounces in ten gallons of water for brining gives a concentration of 1000 parts per million of nitrite, which results in a concentration in the fish of 100 to 200 parts per million. This is the level needed for protection. Make very careful and precise measurements. This is a case where some may be good but more is definitely not better.

*Brined or Dry-Salted Fish.* Fish may be either brined or dry salted before smoking. To brine, mix 32 ounces of salt with each gallon of water. To add flavor to the finished product, include up to 16 ounces of brown sugar, up to ½ ounce of crushed black pepper and a handful of crushed bay leaves in the brine. Soak the fish, keeping it under the brine with nonmetallic weights (as described in Chapter Three) for two to four hours, depending on the thickness and oiliness of the fish and the degree of saltiness required. Thicker and oilier fish require longer salting than thin and lean pieces.

Dry salting is often better, especially for larger fish and large fish cut into chunks. Put a thick layer of salt in a large container. Wash the fish thoroughly, put it on the layer of salt, then sprinkle the wet flesh sides with salt to build up an even layer of about 1/16 inch. Leave the fish in a cool place until the salt is absorbed through the flesh, which may be overnight or longer. You may mix sugar and spices with the salt, as for brining. Do not use more than 1 pound of sugar and ½ ounce of crushed black pepper to 2 pounds of salt.

## Drying

Drying is the second part of the smoking process. This permits the fish to develop a shiny, hard outer crust, called a pellicle. Take the fish from the salt or brine and rinse it briefly in clean, cold water to remove surface salt. Then hang the fish up or spread it on wire screens to dry. Coat screens with vegetable oil to prevent the fish from sticking to the mesh. Do not expose the fish to direct sunlight. Cool, breezy days are best for outdoor drying. Alternatively, use a fan at low speed when working indoors.

Drying is complete when the pellicle forms. This feels dry to the touch and consists of dried, salt-extracted protein, which gives the surface of smoked fish a shiny, attractive appearance. Drying too quickly creates a hard surface that prevents the inner parts of the flesh from drying properly.

Drying takes up to four hours, depending on air temperature, air flow, humidity and the size of the fish. Experience is the best guide.

## Cold Smoking

Cold-smoked products are not cooked and mostly need cooking before they are eaten. Finnan haddie (smoked haddock) is an example. Temperature in the smokehouse ranges from 50°F to 90°F, with the ideal range between 75°F and 85°F.

Smoking times vary from a few hours to as long as 12 hours and may be longer for really large fish. During cold smoking the fish continue to dry. Smoke particles stick to the outside of the flesh, adding flavor and color.

Since fish closest to the smoke source tend to take on a heavier smoke, move them around about halfway through the smoking time. Carp, eels, pollock, kokanee, red drum, walleye and most other firm, white-fleshed fish are excellent cold-smoked.

Smoked salmon is a cold-smoked product that needs no cooking. Because of its great popularity, the following recipe for smoking salmon is included. This is a traditional Scottish recipe, adapted from *A Torry Kiln*

*Operator's Handbook,* published by the U.K. Ministry of Agriculture. (Note that you can also hot-smoke salmon.)

## Smoked Scotch Salmon

Atlantic or chinook (king) salmon are probably the best. Coho (silver) give excellent results, and large chums are also good. Pink and sockeye salmon are less suitable for smoking, though the product is still very good. Because fatter fish are best for smoking, salmon caught in open water before entering their spawning rivers always give a better product than fish that have used up part of their fat-stored energy reserves fighting their way upstream.

Gut the salmon and scrape all traces of blood and kidney from the gut cavity. Head and fillet the fish, leaving the nape bones in place. Thread string through the shoulder of each fillet under the nape bones and tie it in a loop. This gives a convenient handle to use during processing. Score the skin at the thickest part of the fish with a razor or very sharp knife. The cut should just sever the skin, not the flesh underneath. This allows the salt to penetrate more evenly.

Lay the fish, skin side down, on a bed of pure salt about 1 inch thick (see Chapter Three). Add sugar and spices if you desire, but they are not necessary. Cover the flesh side of the fillet with salt about ½ inch thick on the thickest part of the fillet and tapering down to a light sprinkle at the tail. This prevents the thin tail end from becoming too salty and dry.

The salting time depends on the size of the fillet and the fat content of the fish. The fillet loses 8 to 10 percent of its weight when sufficiently salted and feels firm and springy when pressed. If the flesh under the dry outside surface feels soft and fleshy, leave the fish in the salt longer.

A rough guide to salting time: 12 hours for a 1½ to 2-pound fillet; 16-20 hours for a 3- to 4-pound fillet and as much as 24 hours for a 5-pound fillet.

After salting, wash the fillets thoroughly in clean water to remove all the surface salt and hang them to dry for 24 hours in a temperature of 70°F. Then smoke in a very dense smoke for six to seven hours at 70°F to 80°F. Larger fillets may need longer in the smokehouse.

Color is a reasonable guide to smoking progress, but remember that the fish darkens somewhat after smoking is complete. During smoking, expect a weight loss of a further 8 to 10 percent, so that the loss from the starting weight can be as high as 20 percent.

## Hot Smoking

Hot-smoked fish are cooked in hot air, with or without the smoke, to an internal temperature of 180°F (note that this is the temperature at the center of the thickest part of the fish). Hot-smoked fish are ready to eat without further cooking.

Generally, the fish are cold-smoked first for 20 to 60 minutes at about 85°F to 90°F in fairly dense smoke. This dries the fish further and adds to the smoked flavor. Increase the fire or turn up the heaters until the internal temperature of the fish is about 120°F, then increase the heat again to bring the temperature up to about 175°F.

Do not go higher than 180°F, because fish becomes very soft at this high temperature and hanging fish may break from their hangers and fall to the floor. Because the fish are continually cooled by evaporation, the air temperature in the smokehouse will be 10°F to 20°F higher than the temperature of the fish.

At the end of the smoking time the fish should be straw-colored. The color deepens to a golden brown during cooling. Hot-smoked trout and mackerel are both particularly good, as are bonefish, cobia, flounders, catfish, red snapper, grouper and shad.

## After Smoking

Open the smoker and allow the fish to cool to room temperature as rapidly as possible. Hot-smoked fish are fragile until they cool and should not be handled until then.

After cooling, chill both hot-smoked and cold-smoked fish in a refrigerator to as close to $32°F$ as possible. Avoid wrapping smoked fish until it cools since wrapping smoked fish while it is still warm encourages mold growth and early spoilage.

It is better to leave the smoked fish for a day before eating it. This gives time for the salt and smoke flavors to spread evenly through the product and for the flesh to set firm.

The edible shelf life of smoked fish is hardly any longer than that of fresh fish. It varies according to the amount of salt absorbed, the amount of smoke deposited on the fish and the dryness of the product. It is wise to treat smoked product as if it were any other fresh product.

If you do not want to eat it immediately, then freeze it for storage. One week is about the limit for refrigerator storage, assuming that the fish was really fresh when you started smoking it and that the salting and smoking processes were both properly done.

When freezing smoked fish, wrap each piece in aluminum foil or freezer wrap and follow the rules for freezing: spread the fish evenly in a single layer in the freezer; do not put too much in the freezer at one time;

and wrap each piece tightly to exclude as much air as possible.

Remember that freezing tends to highlight the salt flavor in smoked fish. If you expect to freeze some of the product, use a little less salt in the initial cure.

## Liquid Smoke

It is possible to "smoke" fish by using commercially available smoke dips. Used according to directions, these can make an acceptable smoked cod or haddock fillet product. They also add a smoky flavor when used as a baste in the oven or on the barbecue.

## Cod Roe

You can smoke many other marine products. One such delicacy is cod roe. Cod fishermen usually discard the roes, which can be quite large. These are excellent smoked.

Handle cod roes gently or the delicate skins will burst. Keep them well iced up to the point when you start to process them. Wash roes in cold water, then bury them in salt for six to eight hours. After salting, wash them, put them in a wire basket and dip them for 1 minute in boiling water. This plumps them up. Smoke them on trays in dense smoke for six to eight hours at 90-100°F.

You can prepare other large roes in the same way. To use smoked roe, simply slice and sauté it.

You can smoke almost every kind of fish and shellfish, from slabs of giant tuna to butterfish to mussels. Almost all are delicious.

Chapter Six

# One Man's Trash is Another's Treasure

Every angler at some time catches fish he considers to be trash. However, few fish are inedible, and many fish considered by American sport fishermen to be trash are highly regarded by fishermen and consumers in other countries. A poor tasting fish that puts up a good fight is pursued by anglers for its fighting qualities, not for its food value. But most fish are actually good (if not delicious) to eat. Some fish are well regarded in one region and thrown back somewhere else. Put aside the prejudices. Proper handling, preservation and a good recipe can transform almost any fish into a superior dinner for the family. Red horse and other kinds of suckers, buffalofish and freshwater drum or sheepshead are all good to eat. Cusk and wolffish are excellent substitutes for all lean, white-fleshed species such as Atlantic or Pacific cod, as is burbot from the large northern lakes. Atlantic pollock or common carp, both of which have slightly darker meat, can be used similarly. The flavor is excellent and both make a superior chowder.

## Whiting

Whiting are commercially important, but anglers in New England, where the fish is often called silver hake, have traditionally considered them an undesirable species.

However, in the middle Atlantic states, anglers regard whiting as one of the better fish to eat. The flesh is white, sweet and reasonably firm, if you handle the fish properly.

The major disadvantage of whiting is that the flesh can become soft and mushy very quickly if you mishandle it. To prevent this from happening to your catch, gut and bleed the fish immediately after catching it, then wash and ice it as a matter of urgency.

Chilled seawater is an excellent way of preserving fresh whiting at sea. If well iced, it can hold for two or three days without much loss of quality. There is a species similar to Atlantic whiting, called Pacific hake or Pacific whiting, that is found along the West Coast. This also must be handled with speed and care.

You can keep frozen whiting for up to six months at $0°F$. To improve whiting's frozen shelf life, remove the subcutaneous fat that is found underneath the whiting's skin and is about $1/16$ inch thick. After skinning the fillet, slice off this fat layer. This greatly reduces the fat content of the fillet which in turn slows down the development of rancidity while the fish is in the freezer.

Whiting is versatile in use. You can fillet large whiting and substitute it for cod, haddock or other white-fleshed fish in most recipes. Small whiting are usually headed and gutted for pan-frying, baking or broiling.

Smoked whiting is an excellent product, close to lake whitefish. However, do not attempt to smoke Pacific whiting or Pacific hake (species *Merluccius productus*) as it contains an enzyme that causes the flesh to break up during this particular process.

You can use it in other ways—baked, broiled, fried and so on. Many other whiting species can be smoked successfully.

## Red Hake (Mud Hake, Squirrel Hake)

Red hake have rather soft flesh, but are good to eat if you gut, bleed and chill them immediately after landing. Chilled water is recommended for holding this species. The flesh is light colored and makes good fish-and-chips when battered and fried. Do not keep red hake for long in the freezer as the meat gets tough. Two to three months is the maximum life in a home freezer.

## Dogfish

The spiny dogfish and smooth dogfish are members of the shark family and are eaten in many parts of the world. In England dogfish used to be called "rock salmon" and was a popular ingredient in fish-and-chip shops. However, because the name "rock salmon" was considered misleading, the name was prohibited and the fish was renamed "huss," "flake" and eventually "dogfish." These new names are much less appealing than the original "rock salmon" and the fish lost its popularity in England, so proving the marketing power of a name.

The smoked belly flaps of dogfish make the traditional German "Schillerlocken," so called because they curl like hair tresses during smoking.

In New England, many commercial and recreational fishermen consider dogfish a pest. Dogfish migrate into New England waters in the summer at the height of the sport fishing season. They are a nuisance to anglers, stealing bait and often biting and eating hooked fish. It is more than a little frustrating to reel in a fish that is bitten in half. In California, the dogfish and the smoothhound shark, which are both sometimes called grayfish commercially (another unappetizing name), are readily available to anglers.

Dogfish flesh is white, firm and very tasty when fresh. Dogfish, like all other members of the shark family, do not have urinary tracts and excrete body wastes through their skin. Therefore, the major processing problem is that they develop ammonia odors very quickly unless they are well bled, very well iced and processed as soon as possible. Bleed them by cutting the gills and cutting off the tail immediately after catching the fish. Bury dogfish in crushed ice immediately after bleeding them. Figures 25 to 32 demonstrate the processing. The fillets are boneless.

To process dogfish as it is done commercially, first cut open the belly cavity. Then, remove the dorsal fin, starting the cut from the tail and continuing up towards the head (see Figure 25). Then repeat the process by removing the fin on the ventral (belly) side (see Figure 26). Lay the fish on its belly and make a cut behind both sides of the head. This cut should stretch across the initial cut you made to remove the dorsal fin (see Figure 27).

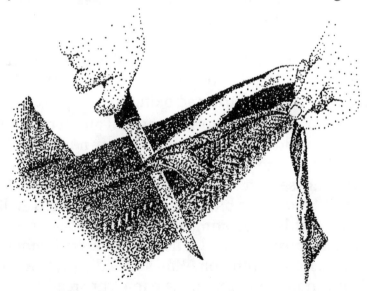

**Figure 25.** Dogfish: removing dorsal fin.

Figure 26. Removing ventral fin.

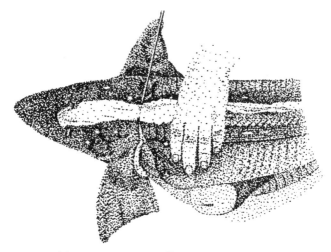

Figure 27. Making cuts across first cut.

Then grab the skin from behind the head and pull it off towards the tail (see Figure 28). This procedure requires a great deal of strength and skill. Because of the roughness of the skin and the way it adheres to the flesh, it is also an unpleasant task. Using a knife or a pair of pliers to pull the skin off makes the job a bit easier. After you pull the skin off, sever the backbone and pull the head

**Figure 28.** Pulling skin off.

and viscera away together (see Figure 29). If you did not remove the tail during bleeding, do so now (see Figure 30). Cut the fillets away from the cartilage backbone (see Figure 31) and you end up with two neat, boneless fillets (see Figure 32).

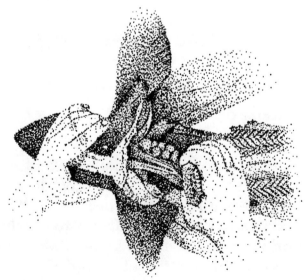

**Figure 29.** Cutting backbone.

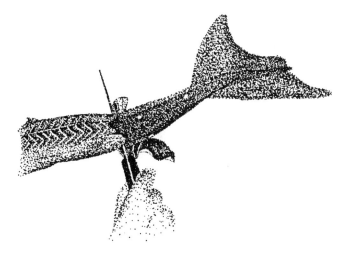

Figure 30.   Removing tail.

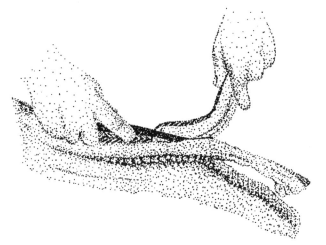

Figure 31.   Cutting fillets.

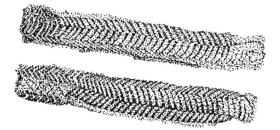

Figure 32.   Dogfish fillets.

Since the angler is not as concerned as the commercial processor with getting every last ounce of meat from the dogfish, it is much easier to fillet the fish as you would any other round fish. You can then pull the skin from each fillet quite easily. Cut between the skin and flesh at the head end to get it started. Grip the skin with pliers and pull it backwards towards the tail, holding the flesh with your other hand. Trim the belly flaps from the fillet. The fillets are boneless because the dogfish has only cartilage, no bones.

If you find that your dogfish has an ammonia smell, soak it in a brine or acid solution (vinegar or lemon juice) to neutralize the odor. You should not have to worry about your dogfish having an ammonia smell as long as you handle it properly.

Do not freeze chunks of dogfish meat for longer than two to three months because of its relatively high fat content. You can fry, broil or bake dogfish.

## Squid (Calamari)

You can catch a large number of squid when jigging for bait or fishing for mackerel. Outside America, especially in the Orient and in southern Europe, squid is widely eaten. It is steadily gaining popularity in the USA as well.

Two species of squid are caught off the northeastern United States. Long-finned squid (known also as winter or bone squid) are found from Cape Cod to Cape Hatteras. The species is *Loligo peallei*. Short-finned squid (known also as summer or illex squid), species *Illex illecebrosus*, are found along the Atlantic coast from Cape Hatteras to Newfoundland. The illex squid is tougher and considered less desirable than the loligo species. It is also more abundant. A smaller loligo species (*Loligo opalescens*) is caught off the West Coast, mainly in California. It is commonly known as market or opal squid. Giant squid are also caught seasonally off California.

Squid is a mollusc, related biologically to clams, scallops, mussels and oysters. Instead of an external shell, it has a light, plastic-like "pen" inside the body. Squid are also fast-moving and voracious predators, unlike their more familiar, shell-encased cousins.

The proportion of edible meat (over 70 percent) in squid is one of the highest of all seafoods. The flesh is white, boneless and very mild. If properly cooked, it is tender and adapts well to many different flavorings.

Chill squid immediately after catching it. Chilled seawater is an excellent method of holding squid at sea. Process squid as soon as possible and then freeze it. Squid freezes very well and stores for over a year at $0°F$. Squid is also one of the few seafoods which is little harmed by defrosting and refreezing several times. Freeze squid whole, or process it first as shown in Figures 33 to 37.

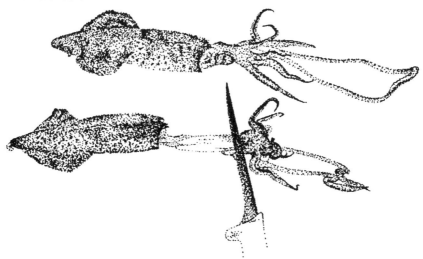

Figure 33.   Squid: cutting tentacles off.

Squid is easy to process, unless you want tubes for stuffing or rings (strips are much easier to prepare and work equally well in recipes that call for rings).
To process into strips:

1. First cut off the head and tentacles (see Figure 33).

2. Then, cut the body lengthwise, spread it out flat (see Figure 34) and scrape away the pen and viscera.

3. The body is now ready to be cut into strips (see Figure 37).

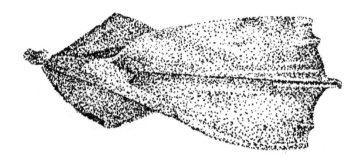

Figure 34.   Cutting the body lengthwise and spreading it flat.

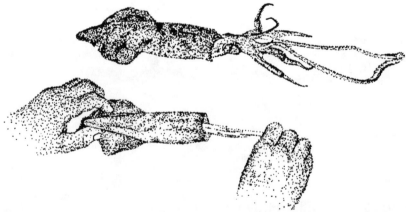

Figure 35.  Pulling the head and tentacles away with the viscera.

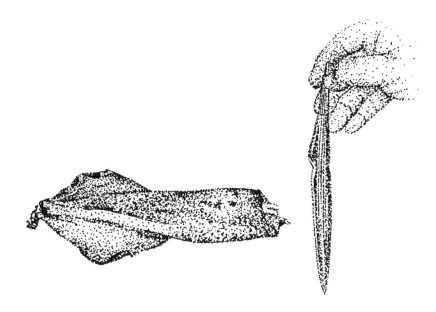

**Figure 36.** Removing the "pen" from the squid.

Making tubes is much harder because the body must be left intact.

To process whole squid into tubes:

1. Hold the head and tentacles and pull them away together with the viscera (see Figure 35).

2. Remove the pen (see Figure 36), which looks like a piece of clear plastic running the length of the body.

3. Clean out as much of the remainder as possible by using your fingers or by squeezing the body tube from the end like a tube of toothpaste.

4. Finally, turn the tube inside out and scrape off any material still remaining. This process sounds quite simple but is actually time-consuming and irritating.

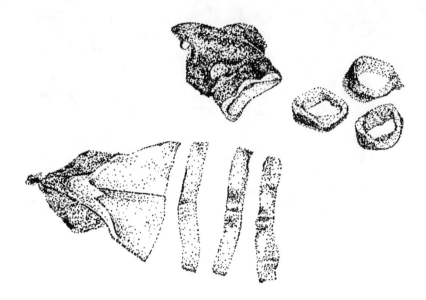

**Figure 37.** Cutting squid into strips or rings.

You now have tubes, which you can turn into rings by slicing across the body vertically into segments about ⅜ inch wide (see Figure 37).

Cut the tentacles away from the head and, if you wish, retain these also. Commercial methods of processing squid are designed to yield the maximum amount possible of edible meat. The instructions given here are slightly less efficient, but are quicker and easier.

Squid have thin skin, which varies in color from white to purple and is often heavily blotched. The skin color bears no relation to the freshness or quality of the fish. It is due to the squid's ability, when alive, to change color to camouflage itself and escape predators. The color left in the skin changes during storage, even in the freezer. You can bleach squid to remove most of its color by soaking it in well-iced seawater for about 30 minutes. In some recipes, for the sake of appearance, you may want to remove the skin. You can do this simply by soaking the fish in hot tap water for one to two minutes, then peeling or rubbing off the skin by hand. If you want squid tubes, you can avoid having to skin the fish by turning the mantles (the tubes) inside out.

Cooking squid is a chef's delight. The bland flesh takes on any desired flavor. You can marinate it, fry it, bake it, add it to sauces and use it in salads. However, you must handle it carefully to keep it tender. Overcooking makes it tough, sometimes very tough. The simple rule is to cook squid either at high temperature ($350°F$) for a very short time (one or two minutes ONLY) or at a very low, simmering, temperature for 30 to 40 minutes. For a spaghetti sauce, cook squid for as long as 60 to 90 minutes. If cooked long enough, squid that toughened in cooking will again become tender.

## Skates and Rays

Barndoor skate, big skate, California skate, bat ray, cownose ray and several other species available all around the USA and often caught by anglers are all good to eat but are commonly killed and thrown back into the sea. This is a waste, especially as skates and rays are easy to clean and prepare.

Handle these fish carefully. Stun them as soon as you catch them. Some species have sharp spines and rough skin that can give you painful wounds and scrapes. To avoid potential problems, wear slip-proof gloves. Remove the spine at the base of the tails of cownose rays. The only edible portions of the skate are the fins, or "wings," so cut these off as soon as possible (see Figure 38) and bury them in ice. You can throw back the rest of the fish or use it as bait.

**Figure 38.** Skate: cutting off wings.

Skates are a unique species in that they develop a sweet, shellfish-like flavor over a period of time. Skate develops its best flavor and texture after two or three days on ice.

The wings have a layer of cartilage running through them horizontally. Fillet them (see Figure 39) and skin

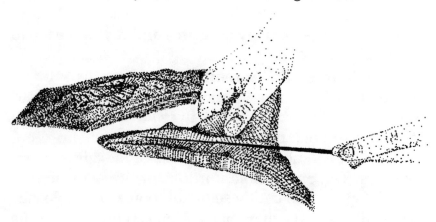

**Figure 39.** Filleting the skate.

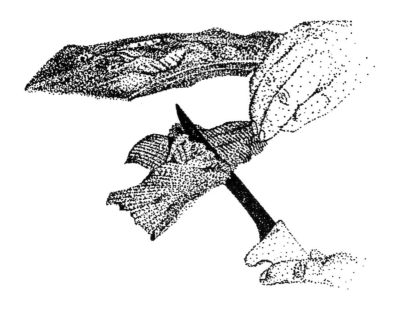

Figure 40. Skinning the skate.

them (see Figure 40) in much the same way as a fish fillet is skinned (see Chapter One). If the wings are small, remove the dark skin and cook the rest of the piece, removing the meat from the cartilage after cooking it. The meat flakes away easily with a fork.

Skate lasts a long time in the freezer. You can use it in a wide variety of seafood dishes and can fry, broil, bake or barbecue it. The texture is more fibrous than that of most bony fish and is more like beef or pork. In fact, the meat from the cownose ray is sufficiently like beef that you can use it instead of beef in stew recipes.

Since fresh skate has a slight shellfish flavor, many people believe that skate wings can be cut to resemble scallops. This is not likely: the work involved would be excessive and the grain of the flesh is different. But the myth indicates how good and palatable the flesh is.

## Anglerfish (Monkfish, Allmouth, Goosefish, Ocean Blowfish, Mother-in-law Fish)

Monkfish, goosefish, anglerfish, or whatever else you may call it, is one of the more grotesque looking fish of the Atlantic coast. Monkfish are known for their voracious eating habits, feeding on almost anything from flounder to lobster to seabirds. Until a decade or so ago, they were discarded at sea, being merely a by-catch of fishing for flounder and sea scallops. Then, American fishermen began to realize that monkfish is considered a gourmet item in many European countries. Monkfish liver is especially popular in Japan, with fishermen receiving five dollars a pound for it.

Now, the species is readily acceptable in the United States, thanks in part to the efforts of Julia Child, who featured monkfish on her famous TV program, "The French Chef." Although they are no more fun to catch than a rock, they are superior seafood. Occasional specimens reach 75 pounds, which gives them some value for the scrapbook and photo album.

The tail is really the only edible body part. To prepare it, cut the tail section just behind the head (see Figure 41). Discard the entire head section. Take care, for the jaws are powerful and can snap dangerously even after the fish is killed and cut. There are several skin layers on the tail section. Remove all of these by pulling them off from the larger end towards the tail fin (see Figure 42). Cut the fish into two fillets by slicing down and around the central bone (see Figure 43). There are no small bones in the tail. The meat is white and firm, and you can use it in almost any recipe calling for white fish. You can also use it as a substitute for lobster meat: steam the pieces, slice them, then serve with drawn butter; or allow them to cool and add them to a lobster salad.

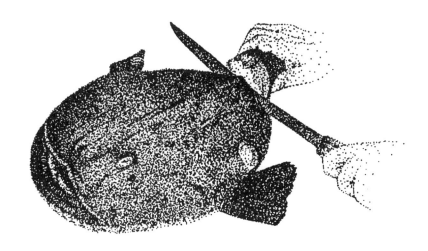

Figure 41.   Monkfish: cutting the tail section.

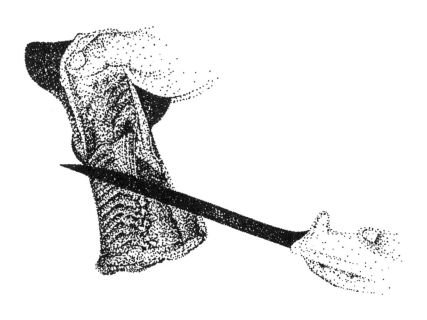

Figure 42.   Removing the skin from a monkfish.

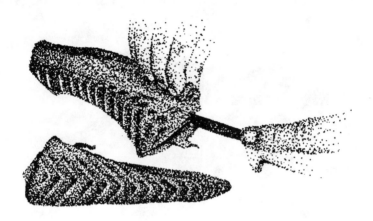

**Figure 43.** Cutting the tail into fillets.

## Sea Robins (Sculpins)

Most anglers consider sea robins to be a pest. They seem to bite at any bait and are difficult to remove from the hook because of their sharp spines. Few fishermen bother to bring them on board, usually sticking them with a knife and cutting them away from the hook.

Several close relatives of the sea robin, called gurnards, are commonly eaten in southern Europe. As with anglerfish, only the tails are usable. Because of the low meat yield, small fish are not worth the effort of saving. Cut off the tail just behind the vent and ice it well for further processing. Fillet and skin it just like a small fish. The flesh is white and firm. In Europe, it is served both hot and cold and is often added to salads.

## Et Cetera

You can put many parts of fish to good use and provide surprisingly palatable food. It costs little or nothing to experiment, anyway.

During the spawning season the fish roe (eggs) are contained in sacs in the body cavity. If you gut the fish carefully, you can remove the roes intact. They are a delicacy pan-fried. Shad, herring, cod, Pacific pollock, bass, walleye, saugers and even flounders all provide good roes. Sturgeon roe, of course, is caviar after salting, and you can treat salmon roe similarly. It is less expensive than caviar and possibly as good. You can also process paddlefish and lake trout roe in the same way. Except for the roes of puffers (blowfish), gar and bowfin, all of which are poisonous, most other roes are palatable. They are usually cooked. Wash them thoroughly and carefully. Then heat them gently in boiling water for a couple of minutes to set them and make them easier to handle. Finally, pan-fry them in butter, bacon fat or olive oil.

You can also pan-fry milts (the male gonads). Herring milts are the best, and many other fish can supply these unusual and tasty parts.

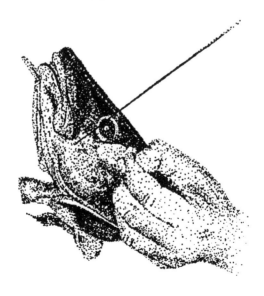

Figure 44. Cutting cheeks from large fish.

Cut cheeks out from large fish such as cod. Figure 44 shows how. You can use these medallions of flesh like scallops. Cod tongues are also edible and are commonly eaten in New England and in the Canadian Maritimes. Push the fish head onto a spike attached to a bench so that the spike penetrates the base of the tongue. Pull the head back and down to expose the base of the tongue, then cut it out (see Figure 45). Deep fry the tongue for best results.

In parts of New England and Canada, fried cod heads are a delicacy. Remove the eyes and gills, split the head down the middle and either pan-fry or deep fry them in breading. Eat the pieces like fried chicken.

Use fish heads and bones from filleting to make stock for soups and chowders.

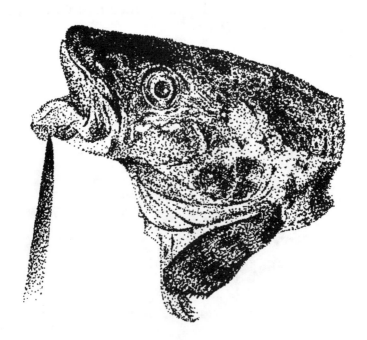

Figure 45. Cutting the tongue from cod.

## Regional Fishes

Most anglers realize that there are some fish that taste much better than others. Yet an angler will go out of his way to catch a fish that tastes awful as long as it puts up a good fight.

Table 5 lists the anglers' favorite regional species of marine fish. It is based upon the listing found in *Current Fishery Statistics,* published each year by the National Marine Fisheries Service. Features and assets of the various fish are presented as well as their common regions.

The amount of fat in the edible portion of the fish is listed under Fat Content. Lean fish range from very little up to 3 percent fat. Moderately fat fish range between 3 and 6 percent. Fat fish may range from 6 to 20 percent fat or even more. The fatter the fish are, the more calories they contain. However, fatter fish also contain more omega-3s. As will be discussed in Chapter Eight, *Fish and Nutrition,* one of fish's greatest virtues is these polyunsaturated fatty acids that can lower the risk of certain forms of heart disease.

The regions identified by number in Table 5 are as follows:

| | | |
|---|---|---|
| 1 | | Atlantic coast from Maine to New York |
| 2 | | Atlantic coast from New Jersey to Cape Hatteras, NC |
| 3 | | Atlantic coast from Cape Hatteras, NC, to Florida Keys |
| 4 | | Gulf coast from Florida Keys to Mississippi River |
| 5 | | Gulf coast from Mississippi River Delta to Mexico |
| 6 | | Pacific coast from Mexico to Point Conception, CA |
| 7 | | Pacific coast from Point Conception, CA, to Washington and Alaska |

### Table 5: Regional Fishes—Their Sensory and Nutritional Potential

| Species | Region | Flavor | Color Cooked | Fat Content | Remarks |
|---|---|---|---|---|---|
| Albacore | 3,6,7 | fairly mild | light | fat | Most valuable and desirable tuna |
| Amberjack (greater) | 3,4,5 | strong | mod. dark | fat | Adult fish often wormy |
| Barracuda (great) | 3,4,5 | avoid | light | lean | Adult fish can be toxic |
| Barracuda Pacific | 6 | fairly mild | light | mod. fat | Good eating, not toxic |
| Bass, black sea (Atlantic) | 2,3 | mild | white | lean | Tackle record 8 lb. |
| Bass, giant sea, Pacific | 6 | mild | light | lean | Can reach 7 feet and 500 lb. |
| Bass, kelp | 6 | mild | light | lean | Desirable game species |
| Bass, striped | 1,2,3,4, 6,7 | fairly mild | light | lean | Big fish, somewhat coarse flesh |
| Bluefish | 1,2,3,4 | strong | dark | mod. fat | Tackle record 31 lb., 12 oz. |
| Bocaccio | 6,7 | mild | fairly light | lean | Important sport and commercial fish |
| Bonefish | 3,4 | strong | light | lean | Excellent smoked |
| Bonito (Atlantic) | 1,2,3 | fairly strong | mod. dark | mod. fat | Best commercial canning |
| Bonito (Pacific) | 6 | fairly strong | dark | fat | |

Table 5: Regional Fishes—Their Sensory and Nutritional Potential (Continued)

| Species | Region | Flavor | Color Cooked | Fat Content | Remarks |
|---|---|---|---|---|---|
| Cobia | 3,4,5 | mod. strong | white | mod. fat | Excellent smoked |
| Cod, Atlantic | 1,2 | mild | white | lean | Very versatile fish |
| Cod, Pacific | 7 | mild | white | lean | Very versatile fish |
| Croaker, Atlantic | 2,3,4,5 | fairly mild | light | lean | Small size, good eating |
| Drum, black | 2,3,4,5 | fairly mild | white | lean | Tackle record 98.5 lb. |
| Drum, red | 2,3,4,5 | fairly mild | light | lean | Adults often wormy flesh |
| Flounder, summer | 1,2 | mild | white | lean | A flounder at its finest |
| Flounder, winter | 1,2 | mild | light | lean | Valuable commercial and sport fish |
| Grouper, black | 3,4 | mild | light | lean | Good fighter, good eating |
| Grouper, red | 3,4 | mild | light | lean | Common and excellent eating |
| Grunt, white | 3 | mild | white | lean | Good pan fish in Florida Keys |
| Haddock | 1 | mild | white | lean | The famous finnan haddie when smoked |
| Hake, silver | 1,2 | mild | light | lean | Excellent smoked |

(Cont'd)

Table 5: Regional Fishes—Their Sensory and Nutritional Potential (Continued)

| Species | Region | Flavor | Color Cooked | Fat Content | Remarks |
|---|---|---|---|---|---|
| Halibut, Atlantic | 1 | mild | | mod. fat | Expensive but excellent |
| Halibut, Pacific | 7 | mild | white | lean | Expensive but excellent |
| Jack, crevalle | 3,4,5 | fairly mild | fairly light | lean | Hard fighter, poor food |
| Ladyfish | 3,4 | fair | light | lean | Seldom eaten, many fine bones |
| Lingcod | 6,7 | mild | light | lean | Flesh at times green but good |
| Mackerel, Atlantic | 1,2,3 | strong | dark | fat | Over 3 lb. is unusual |
| Mackerel, chub | 2,3,6,7 | strong | dark | fat | Over 1 lb. is unusual |
| Mackerel, king | 2,3,4,5 | strong | dark | mod. fat | Tackle record 78 lb. |
| Mackerel, Spanish | 3,4 | strong | dark | fat | Over 4 lb. is unusual |
| Mahi-Mahi | 2,3,4, 5,6,7 | fairly strong | white | lean | One of best game fish |
| Marlin, blue | 3,5 | fairly mild | fairly light | mod. fat | Good food, females grow much larger |
| Marlin, striped | 5 | fairly mild | light | lean | Esteemed food in Orient, not U.S. |
| Mullet, striped | 3,4,5 | fairly mild | light | fat | Atlantic, good; Gulf, oily |

Table 5: Regional Fishes—Their Sensory and Nutritional Potential (Continued)

| Species | Region | Flavor | Color Cooked | Fat Content | Remarks |
|---|---|---|---|---|---|
| Pollock | 1 | mild | fairly light | lean | Sportiest fish of cod family |
| Pompano, Florida | 3,4,5 | fairly mild | light | mod. fat | Epicure's delight |
| Puffer, northern | 2,3 | mild | white | lean | Liver can be toxic |
| Rockfish, chilipepper | 6,7 | fairly mild | light | lean | Best from central and southern California |
| Rockfish, yellowtail | 6,7 | fairly mild | light | lean | Good game fish off reefs |
| Sablefish | 6,7 | mild | white | fat | Top smoked delicatessen fish |
| Sailfish | 3,4,5 | strong | light | lean | Seldom eaten |
| Salmon, chinook | 6,7 | fairly strong | fairly dark | fat | Highly desirable salmon |
| Salmon, coho | 6,7 | fairly strong | fairly dark | mod. fat | Can grow to 30 lb., average under 10 lb. |
| Salmon, pink | 6,7 | fairly mild | fairly dark | mod. fat | When adult, average 5 to 6 lb. |
| Sea trout, sand | 4,5 | mod. strong | light | mod. fat | Popular Gulf fish, flesh soft |
| Sea trout, spotted | 3 | fairly mild | white | lean | Good fish; flesh firmer |
| Shad, American | 1,2,3 | fairly mild | mod. dark | fat | Tasty but bony; flesh soft |

(Cont'd)

Table 5: Regional Fishes—Their Sensory and Nutritional Potential (Continued)

| Species | Region | Flavor | Color Cooked | Fat Content | Remarks |
|---|---|---|---|---|---|
| Shark, porbeagle | 1,2,3 | fairly strong | light | lean | Swordfish substitute |
| Shark, silky | 3 | fairly strong | fairly dark | mod. fat | Valuable hide |
| Shark, thresher | 3,4,6 | fairly mild | light | lean | Good food, excellent smoked |
| Snapper, red | 3,4,5 | mild | white | lean | Highly prized |
| Snapper, yellowtail | 3,4 | mild | white | lean | Highly prized |
| Sole, English | 6,7 | fairly mild | white | lean | Good, but has medicinal flavor at times |
| Sole, Petrale | 6,7 | mild | white | lean | Among best of flatfishes |
| Spot | 2,3 | fairly strong | mod. dark | lean | Flavor good; flesh soft, sometimes wormy |
| Swordfish | 6 | mod. strong | light | mod. fat | Excellent and expensive |
| Tautog | 1,2 | fairly mild | light | lean | Good angling, good eating |
| Tilefish | 1,2,4,5 | mild | light | lean | Good eating |
| Tuna, bluefin | 1,2,3,6 | fairly mild | dark | fat | Fattest of tunas |
| Tuna, skipjack | 2,3 | fairly strong | mod. light | mod. fat | High economic value |
| Tuna, yellowfin | 2,3,6 | fairly mild | mod. light | mod. fat | High economic value |

Table 5: Regional Fishes—Their Sensory and Nutritional Potential (Continued)

| Species | Region | Flavor | Color Cooked | Fat Content | Remarks |
|---|---|---|---|---|---|
| Wahoo | 2,3 | fairly mild | white | lean | Good fighter, good eating |
| Weakfish | 2,3 | mild | white | mod. fat | Prized sport fish, flesh soft |
| Wolffish | 1 | mild | white | lean | Beware of their powerful jaws |
| Yellowtail | 6 | fairly strong | mod. light | lean | Popular sport fish |

*Chapter Seven*

# Yes, Virginia, Seafood is Safe to Eat

The safety of the fish we eat has become a major public issue, with much heat and little light generated by the poorly informed comments of consumer interest groups and parts of the press. There is a strong political movement towards mandatory inspection of all seafoods to ensure that consumers are fully protected.

The facts are really not alarming at all. Indeed, a 1988 report from the federal government's General Accounting Office concluded that incidents of illness associated with seafoods were so rare that "there does not appear to be a compelling case at this time for implementing a comprehensive, mandatory federal seafood inspection program similar to that used for meat and poultry."

The angler should note further that the vast majority of cases of illness associated with seafoods were blamed on contaminated shellfish, not on finfish. Even the world's most innovative angler could not persuade an oyster to snap at a baited hook (although mussels and clams are occasionally caught this way). And many of the cases involving contaminated shellfish were the results of people disregarding harvesting laws and restrictions (whether willfully or unknowingly, it makes no difference to the pain in your gut if you are a victim). The greater part of the remaining cases were ciguatera on reef fish in the Caribbean and Hawaii and scombroid poisoning from improper handling.

What all this means to the angler is that the fish you catch is good and safe to eat almost all the time.

The rest of this chapter describes the various hazards that do exist. Bear in mind that the risks are small, and common sense helps to reduce them further. Follow local guidelines: if there are restrictions on fishing, obey them. Ice your fish as soon as you catch it. Don't let it start to decay, which adds to the risk of histamine poisoning. Fish tastes much better fresh, anyway.

## Poisoning Due to Biotoxins

### Tetrodotoxin Poisoning

Also known as puffer-fish poisoning, tetrodotoxin poisoning results from eating puffer fish (globefish or blowfish), porcupine fish and ocean sunfish. Triggerfish and filefish are also suspect. The problem is serious in Japan, where it is called fugu poisoning ("fugu" is the Japanese name for puffer fish). In response to the danger, the federal Food and Drug Administration banned imports of puffer fish. Imported species are much more toxic than the domestic ones. However, the FDA recently allowed Japanese fugu back into the United States under strict inspection. This response is due to the intense demand for fugu sashimi in Japanese restaurants. Puffer fish from the northeast coast of the United States are considered harmless as long as the flesh is completely cleaned of visceral materials and skin. Cleaned Atlantic puffer fish are sold as "sea squabs."

The toxin, which is resistant to heat, is present in the reproductive organs, the viscera (particularly the liver) and the skin. The origin of the tetrodotoxin, as the poison is called, remains a mystery, although toxic dinoflagellate algae (minute planktonic creatures) are suspected. The symptoms of this type of intoxicant include a tingling of the lips, mouth and extremities, followed by numbness in the limbs and general loss of muscular coordination. Symptoms appear soon after eating the poison, and death may follow within 2 to 12 hours. The mortality rate can be as high as 50 to 60 percent. In Japan, the problem is controlled by permitting puffer fish to be sold only in specially licensed restaurants where the chefs are required to prove both their knowledge of the hazard and their proficiency in safely cleaning and preparing the fish for cooking without rupturing any of the toxic organs. Although puffer fish along the eastern U.S. coast are not necessarily poisonous, sport fishermen should be aware of the potential danger.

**Ciguatera Poisoning**

More than 300 different fish species found in tropical and subtropical waters are occasionally responsible for ciguatera poisoning. The Florida/Caribbean region and the Hawaiian Islands are the areas of major concern in the United States. Barracuda, jacks, snappers, groupers, sea bass, parrot fish and surgeon fish are the species most often involved. Fish associated with this illness may be perfectly edible when taken from some areas, but toxic when taken from others. The causative agent has not yet been positively identified; it is thought to be one of the dinoflagellates consumed by fish indirectly through the food chain. Large fish are more likely to contain the poison than are small fish of the same species, probably because the larger fish have more opportunity to accumulate the toxin. Do not eat barracudas from the Florida area that are larger than three pounds. The smaller barracudas are relatively free of the poison. The reason for this is probably that the small barracudas feed close to shore, whereas larger fish feed offshore where they can prey on other ciguatoxic fish and accumulate toxic levels of the poison.

The symptoms of ciguatera poisoning include gastrointestinal disturbances (vomiting, cramps, diarrhea), aching of the muscles and joints, extreme weakness, tingling sensations, inverse temperature sensitivity (hot seems cold and cold seems hot) and intense itching. Symptoms can occur soon after eating and are usually experienced within six hours. The onset and severity of the symptoms are related to the amount of toxin eaten. The toxin is heat resistant, so is not destroyed by cooking.

Toxic fish display no abnormal signs and there is no simple way to differentiate between safe and unsafe fish. The mouse bioassay is the only sure test. In a mouse bioassay, an extract of the fish flesh is injected into mice to test its deadliness. A lethal dose is one required to kill a 20-gram mouse in 15 minutes. The minimum human lethal dose is estimated to be some 3000 to 5000 mouse units (M.U.) and a dose of 600 M.U. can produce symptoms.

There is no known antidote for the tetrodotoxin poison and the only recourse is to treat the symptoms under a physician's care. The best course of action is to flush the gastrointestinal tract at the earliest sign of intoxication. The safest way to prevent ciguatera poisoning is to avoid eating any types of large fish that have a history of being implicated.

## Scombroid Poisoning (Saurine Poisoning)

Fish such as tuna, bonito, mackerel-type fish, swordfish, mahi-mahi and bluefish, whose flesh is normally nontoxic, can become toxic when exposed to warm temperatures for several hours after catching. Never leave these species out on deck, especially during hot temperatures. Under these conditions, certain bacteria present on the fish may reproduce rapidly and convert histidine, a naturally occurring amino acid in the flesh, into histamine or histamine-like compounds, which cause allergic reactions in humans. Symptoms include headache, dizziness, burning sensation, gastrointestinal disturbance, abdominal pain and skin eruptions. Taking antihistamines may give relief. The scombrotoxin is heat resistant and outbreaks have occurred from eating commercially canned tuna. The best preventive measure is to ice or refrigerate the catch immediately after landing. Do not eat fish with a tainted or rancid flavor, which may indicate considerable bacterial activity.

**Paralytic Shellfish Poisoning (PSP)**

Anglers collecting shellfish for bait and for their own consumption should be aware of the importance of obeying harvesting restrictions. These are designed to protect the public from shellfish contaminated with viruses, bacteria and "red tide."

Eating shellfish that contain a toxin produced by their ingestion of poisonous dinoflagellates causes the illness known as paralytic shellfish poisoning, or PSP. The causative organisms are: *Gonyaulax catanella* on the West Coast, where it is found from Alaska to Washington; *Gymnodinium breve* in the Florida area, and *Gonyaulax tamarensis* along the northeast coast ranging from the Canadian Maritimes to Cape Cod.

Molluscs, including mussels, clams, quahogs, oysters, scallops and whelks, accumulate the toxin in their digestive glands. Some clams also carry the poison in their siphons, or "necks." The shellfish do not appear to be harmed by the presence of the toxin. Unfortunately, one cannot tell merely by looking whether a particular batch of shellfish is safe to eat.

Mussels are toxic to eat when the concentration of the causative algae in the seawater is about 200 cells per milliliter, roughly 200,000 per quart. When the cell density reaches about 1000 per milliliter, roughly one million per quart, the seawater takes on a reddish color, a phenomenon often called "red tide."

To protect public health, the cell count of the plankton in shellfish waters is monitored regularly to ensure a safe level. If this is exceeded, the shellfish beds are quarantined and harvesting from them prohibited. This is one of a number of reasons why ignoring closed-area rules is unwise.

The symptoms of PSP appear soon after consuming the contaminated shellfish. They begin with a tingling of the lips, mouth and extremities, followed by numbness in the arms and legs and general loss of muscular coordination. As the illness progresses, breathing becomes more difficult. Death within 2 to 12 hours may result from paralysis, depending upon the amount of toxin the person takes in and the person's body weight.

Usually, if the victim survives the first 24 hours, an uneventful recovery follows. The minimum human lethal dose is estimated to be 3000 to 5000 M.U. (mouse units, see above). A dose of 600 M.U. can produce symptoms.

The poison, called saxitoxin or brevitoxin, acts on the human nervous system similarly to curare. It is one of the most potent toxins known to man and there is no known antidote.

A victim should induce vomiting and then take a purgative or an enema to clean out the gastrointestinal tract as soon as symptoms appear. Cooking may destroy a small part of the toxin, but it does not make the shellfish safe. Even after heating in a pressure cooker to 240°F, toxic clams still contain about half of the poison.

A fisherman gathering shellfish in an unmarked area should learn to recognize warnings of red tide. Discolored sea water is an obvious sign. Dead seabirds along the shore can also be an indicator since they may have eaten the poisonous shellfish.

Do not, under any circumstances, use shellfish from areas closed to shellfishing. Shellfish may remain toxic for several weeks after the red tide has passed. Never eat shellfish purchased for bait. Bait dealers will often buy clams that are unfit for human consumption.

Toxic outbreaks of PSP are sporadic. Its occurrence is most likely from June through October. Heavy algal blooms are fostered by a combination of warm temperature, low salinity, an abundance of nutrients and low levels of certain trace metals (particularly copper).

Mussels are the most likely mollusc to accumulate the toxin. They contain the highest toxin levels and are usually the first to decontaminate themselves. PSP was sometimes called "mussel poisoning" because of the frequency with which mussels were involved.

Although fish are not generally affected by the presence of toxic algae in the water, they may die if they eat the toxic shellfish. Large fish kills are sometimes observed in red tide areas, but these deaths are probably due to oxygen depletion by the large masses of algae.

A recent discovery by the National Marine Fisheries Service's Gloucester Laboratory shows that the livers and viscera of Atlantic mackerel contain levels of PSP toxin ranging from 0 to 100 M.U. An extensive two-year survey shows that only the mackerel livers contain toxin and that the edible flesh and roe show negative toxicity.

## Poisoning Due to Chemicals

### Chlorinated Hydrocarbons

Polychlorinated hydrocarbons have been widely used industrially in past years. Some of them, like DDT, Dieldrin and Kepone have been effective insecticides; another, a polychlorinated biphenyl (PCB), is primarily used as a coolant insulator in electrical transformers and capacitors. These chemicals have contaminated major inland waterways in the United States, including the Great Lakes and the Mississippi and Hudson rivers and many other waters. They are toxic to man and animals, perhaps even cancer-causing. They are very stable and do not decompose readily into less complex and less harmful substances.

Fish may contain high levels of PCBs without being overtly affected, except that the presence of PCBs may impair reproduction. Shrimp and shellfish are extremely sensitive to PCBs and may die even when only small concentrations are present in the water. The chemicals get into fish primarily through the food chain but also by absorption through their gills or through the skin. Larger carnivorous fish usually contain higher levels of PCBs than do smaller fish. Fish with higher fat content will usually display higher levels of PCBs than lean fish.

Commercially harvested fish present less of a hazard since they are mostly caught offshore in deep water where the concentration of chemical contaminants is low. The problem seems to be confined principally to some large fish caught in certain areas of the Great Lakes and major estuaries. A survey of several saltwater species revealed that PCBs, when present, are at safe low levels, on the average below the 2 parts per million limit established by the federal FDA. Some individual fish are occasionally found with PCB concentration exceeding the federal limit of 2 parts per million.

The fat of contaminated fish contains ten times more PCBs than the lean muscle. Any hazard from eating them greatly diminishes by removing as much as possible the fat (the dark muscle) and the fatty strip along the lateral line or back section. Discard the skin and belly flaps, which are both usually rich in fat. Use cooking methods that render out much of the fat, such as broiling, baking or barbecuing. Avoid using recipes and cooking methods where you cook the fish in its own juices, such as chowders and pan-frying. Try not to eat suspect fish frequently and avoid the larger specimens altogether. Pregnant women especially should limit their consumption of these. Fortunately, this hazard occurs with few species.

**Heavy Metals**

Owing to their widespread distribution throughout soils, lakes, rivers, oceans and even air, traces of heavy metals are found in virtually all plants and animals. Lead, cadmium and mercury, all of which are toxic at high levels, are of particular concern. Heavy-metal poisoning in humans can result in neurological disturbances, congenital defects at birth and even death. Worldwide interest in the presence of these toxic metals in the food supply was stimulated by an outbreak of mercury poisoning in Japan in the late 1950s, resulting from the consumption of heavily contaminated fish taken from Minamata Bay. Marine and fresh waters normally contain a low natural level of mercury. High concentrations may exist in certain lakes and bays as a result of geological influences and pollution from various sources. In the Minamata Bay incident, a plastics plant was discharging mercury-containing waste into a river leading into the bay.

Metallic mercury is converted to methylmercury, a highly toxic form, by bacteria in bottom sediments. The methylmercury is assimilated by plankton, which in turn is taken up by fish and shellfish as their food. The older, larger, predatory fish such as tuna, swordfish, shark, northern pike, walleye and chinook and coho salmon are potentially the most hazardous. Surveys of the flesh of bottom-dwelling marine fishes caught in the northwest Atlantic, such as codfish, haddock and flatfish, have found them to be safe. High concentrations of mercury are most likely to occur in fish living near areas where there are or have been local discharges of mercury-containing wastes. Some shellfish and crustaceans customarily concentrate heavy metals internally. The federal Food and Drug Administration has imposed a public health safety limit of 1 part per million (1 ppm) of methylmercury in the flesh of seafoods. A National Marine Fisheries Service report in 1978 concluded that "Mercury in seafood poses little hazard to the overall seafood eating public."

Table 6 is the FDA's list of harmful substances and their action levels that may be found in fish.

Selenium in the diet may diminish the toxicity of ingested mercury and may possibly play a part in cancer prevention. Selenium is obtained from seafoods, especially shellfish and tunas, as well as from kidneys, grains, beans and nuts.

## Parasitic Infections

Parasites occur frequently on the surface of fish, as well as within the flesh and internal organs. Some are large enough to be seen with the unaided eye. Many parasites are so small they are only visible with a microscope. Surface or external parasites include leeches, fish lice and copepods. The "buttons" that are sometimes noticeable on ocean perch, bluegills and largemouth bass are parasitic copepods, called "anchorworms," which have an anchorlike projection of the head growing into the flesh of the fish. Internal parasites include single-celled organisms (protozoans—myxosporidia, microsporidia), flatworms (trematodes), tapeworms (cestodes) and roundworms (nematodes).

Table 6: FDA Action Levels for Poisonous or Deleterious Substances in Seafood

| | | |
|---|---|---|
| Aldrin | (fish and shellfish) | 0.3 ppm |
| Dieldrin | (fish and shellfish) | 0.3 ppm |
| Benzene Hexachloride | (frogs legs) | 0.5 ppm |
| Chlordane | (fish) | 0.3 ppm |
| DDT, DDE, TDE | (fish) | 5.0 ppm |
| Endrin | (fish and shellfish) | 0.3 ppm |
| Heptachlor | (fish and shellfish) | 0.3 ppm |
| Kepone | (crabmeat) | 0.4 ppm |
| Kepone | (fish and shellfish) | 0.3 ppm |
| Mercury (measured as methylmercury) | (fish, shellfish and crustaceans) | 1.0 ppm |
| Mirex | (fish) | 0.1 ppm |
| PCB | (fish) | 2.0 ppm |
| Toxaphene | (fish) | 5.0 ppm |
| Paralytic Shellfish Toxin | (clams) | 80mg/100gm meat |
| Paralytic Shellfish Toxin | (mussels) | 80mg/100gm meat |
| Paralytic Shellfish Toxin | (oysters) | 80mg/100gm meat |

Note: *ppm* means parts per million, *mg* means micrograms.
Source: U.S. Food and Drug Administration.

Common in both largemouth and smallmouth bass is the bass tapeworm, which occurs in the viscera in larval form, sometimes in great abundance. The fish is edible once you remove its insides. Yellow grubs and black grubs (both trematodes) occur frequently on the skin of freshwater fishes like yellow perch, sunfish and bass, especially near the tail. Once again, the fish is edible once you remove the grubs with a knife. Fish can tolerate the presence of moderate numbers of parasites without apparent harm.

Because of their unappetizing appearance, together with a general uncertainty as to whether parasites pose a health hazard to humans, most anglers discard obviously parasitized fish. Avoid such wasteful discard simply by cutting out the infected portion to remove the parasites. Only a very few fish parasites are known to affect humans.

Some parasitic infections of fish can impair the quality of the flesh. Myxosporidian parasites, though microscopic in size, form black clusters, giving a peppered appearance to the flesh. They are common in yellow perch in freshwater. In addition, these organisms secrete a powerful protein-digesting enzyme that turns the flesh mushy or jelly-like. This applies to halibut, certain hake, swordfish and some other fish. These protozoan parasites can also produce small white nodules ("pus pockets") in the flesh of herring, alewives and menhaden found along the eastern United States coast.

Two fish parasites that can affect man occur in US waters. One is the broad fish tapeworm (*Diphyllobothrium latum*) found in some freshwater fish (northern pike, yellow perch). Although this parasite is not known to infest saltwater fish, it was recently found in Alaskan salmon in freshwater streams. Symptoms of diphyllobothriasis include abdominal discomfort, diarrhea or constipation. The other is a roundworm (*Anisakis simplex*), which causes an infection called "anisakiasis." Herring and mackerel are among the fishes in the northeast United States that can be infested with these worms. Humans are infected by eating raw or undercooked fish. Lightly cured or slightly salted herring are also involved in most cases. The anisakis larvae do not mature and reproduce in the human gut, but they can cause inflammation and ulceration.

Cod flesh can be parasitized with a roundworm (*Phocanema decipiens*) that belongs to the same family as anisakis. These cod worms may also occur in smelt, haddock, plaice, ocean perch and whiting.

Cod worms are more likely in larger specimens of a species and also in fish that inhabit shore areas where seals reside. The worm needs harbor seals as a host to complete its life cycle. Recent evidence indicates that cod worms can cause gastric disturbances in humans and, although rare, can burrow through the intestinal wall, but with much less virulence than anisakis.

## Preventing Parasitic Infections

The best policy for preventing parasitic infections is to avoid eating raw or undercooked fish. Freezing fish for 72 hours or longer at $0°F$ will kill larval worms. Hot smoking of fish usually produces an internal temperature high enough to destroy the parasites. Cold smoking, in which the smoking temperature does not exceed $100°F$, is definitely inadequate for destroying the parasites. However, freezing cold-smoked fish afterwards ensures killing the parasites and does not cause a major loss in quality. Some parasites can survive in brine for a month; therefore, lightly salted or cured fish can pose a hazard.

A common commercial practice for locating parasitic worms in a fillet is "candling." The processor places the fillet on a glass surface above a strong light. This enables him to see the outline of the parasite, which he then removes.

Parasites frequently concentrate in the internal organs but they are removed during cleaning. Belly flaps can become heavily infested in worm-infested fish and should therefore be discarded. Larval worms may migrate from the viscera into the flesh of whole fish held refrigerated for several days. This is an additional reason to gut your fish as soon as possible after it is caught.

These precautions against parasitism are not intended to frighten or discourage anglers from eating their catches. Actually, very few cases of parasitic infection have ever been documented from consumption of fish taken from marine or fresh waters.

## Bacterial Food Infections or Intoxicants

A bacterial food intoxication is an illness derived from the consumption of food in which a harmful bacterium has already multiplied to large numbers and has produced a toxin. Symptoms appear soon after eating. In bacterial food infection, symptoms do not appear until the bacteria multiply in the gut to a critical population. The most prominent symptom in both cases is gastrointestinal upset.

### Clostridium Botulinum and Staphylococcus Aureus

The prime agents of bacterial intoxication in seafoods are *Clostridium botulinum* and *Staphylococcus aureus*.

***Clostridium botulinum.*** *Clostridium botulinum* causes the dreaded and often fatal disease called botulism, which is characterized by muscular and respiratory paralysis. This bacterium is present everywhere in soil, in the marine environment and sometimes in seafoods, though generally in small, harmless numbers. If you give the bacteria the opportunity to multiply, for example when storage temperature is too high, food can become toxic within several hours.

Fortunately, *Clostridium botulinum* can grow only in anaerobic conditions (without air). Since there is such a low incidence of this organism in marine species, the normal spoilage bacteria (aerobic) present in fish spoil the fish before the botulism toxin can form. Botulism toxin has not been found in normally packaged fresh and frozen fish.

Botulism is a potential hazard with smoked fish stored at relatively warm temperatures and with improperly processed canned food. It is fortunate that the toxin is sensitive to heat. Ordinary cooking methods destroy the botulism toxin.

Most fatalities have occurred when contaminated foods such as improperly processed canned tuna and vacuum-packed smoked fish were eaten without being heated.

***Staphylococcus aureus.*** *Staphylococcus aureus* is normally absent from the ocean environment and from seafoods. Its presence on fish results from contamination by human handling. The bacterium is usually present on human skin, hands and the nose. It is frequently found in boils, carbuncles and cuts. People having any of these skin conditions should not handle food. The toxin survives cooking and, though not fatal, can produce highly unpleasant symptoms (vomiting, cramps, diarrhea).

The principal bacterial food infections transmitted from seafoods are caused by *Vibrio parahaemolyticus*, salmonella and shigella.

### Vibrio parahaemolyticus

*Vibrio parahaemolyticus* appears to be a normal inhabitant of the marine environment, including coastal waters of the United States in the warmer water regions. It frequently contaminates fish, molluscs and crustaceans. Warm temperatures accelerate growth, so the greatest threat is during summer months and improper refrigeration.

The bacterium is destroyed by cooking. Symptoms include diarrhea, cramps, vomiting and fever, which usually occur about 15 hours after ingestion. Another related but more virulent bacterium, *Vibrio cholerae*, is a normal inhabitant of estuarine waters, particularly in the southern coastal states. This organism causes cholera. Think about it before you eat raw seafoods.

### Salmonella and Shigella

Salmonella and shigella bacteria are usually found in coastal or estuarine waters polluted by the discharge of raw sewage or of sewage effluent. Bivalve molluscs such as clams, mussels and oysters concentrate these bacteria in their digestive organs. Infection may result from eating raw or insufficiently cooked molluscs harvested from polluted areas. Finfish in polluted waters frequently harbor these microorganisms in their intestines or on their skin, with no apparent ill effects.

Food poisoning by salmonella (salmonellosis) causes gastroenteritis, which is usually accompanied by fever. The disease caused by shigella bacteria (shigellosis) is a rather severe type of dysentery.

## Minimizing Bacterial Risks: A Summary

To minimize bacterial risks from seafoods:

- Do not harvest fish from polluted waters.

- Do not wash fish with polluted water.

- Maintain proper temperature control of the seafood at all times and ice fish immediately. Disease-causing bacteria multiply between 40°F and 140°F. Do not hold seafoods within this temperature range.

- Do not handle or prepare food for eating when you have sores or open cuts, or are suffering from intestinal disturbance.

- Do not prepare seafoods, especially salads, much in advance of serving. Avoid storage at room temperature, which provides favorable conditions for bacteria to grow.

- Do not eat raw seafoods. Cook to an internal temperature of at least 160°F.

## Virus Infections

Shellfish harvested from polluted waters can transmit viral diseases. One particularly nasty example is infectious hepatitis. The best protection is to abstain from eating raw shellfish from suspect areas.

## Radioactivity

At present there is no recorded occurrence of radioactivity in seafoods greater than normal background levels.

## Allergies

Some seafoods contain agents known as "allergens" or "antigens." These stimulate the production of specific antibodies creating sensitivity to that seafood. Subsequent meals of the same seafood induce an allergic reaction and, in a severe case, anaphylactic shock. Symptoms include hives, headaches, swelling and gastrointestinal disturbance.

Some people seem to inherit a predisposition for sensitivity to seafood, but they are usually allergic to only one species or a related group. Fish proteins act as allergens in these cases. Sardines, salmon, mackerel, crabs and lobsters are some of the more frequently implicated species. Seafoods are not unusual in this respect. Allergic responses to many other types of food are well known.

## Bites and Stings

Many aquatic animals, especially in warm and tropical waters, can inflict venomous stings. They include certain corals, cone shells, sea urchins, sea anemones, jellyfish and some fish. Many of these animals have tentacles equipped with stinging cells. Symptoms of stings vary with the individual and the site of the sting. They range from a burning, stinging sensation to throbbing pain accompanied by a localized rash or swelling. Most jellyfish cause some degree of skin irritation and you should therefore avoid all of them. Some fishermen have reported obtaining relief from the stings by applying a paste made from meat tenderizer.

Many species of fish possess venomous spines that can cause painful stings, sometimes even death. Dogfish have spines at the forward base of each of the two dorsal fins. Catfish have spines on the dorsal and pectoral fins.

Stingrays have stingers in their tail. The common skate lacks a tail spine, but the cownose ray and the ship tail stingray both have them. The brier skates, or thorny skates, are armed with a row of thorns along the midline of the back and tail. These rays usually burrow in the mud, which makes them difficult to see. Most injuries from rays occur when people wading barefoot inadvertently step on them. Wash a stingray wound with cold saltwater to remove any venom or pieces of stinger. If stung at an extremity, soak the wound for 30 to 60 minutes in water as hot as you can tolerate. You may add epsom salts to the water. This procedure helps to inactivate the toxin. If the swelling and pain do not subside within a couple of hours, consult a physician for possible suturing and application of antibiotics.

Handle catfish, dogfish and stingrays with extreme caution. Bear in mind that, in addition to a painful sting, the spine can produce severe lacerations. The rough hides of sharks and dogfish can also cause skin abrasions.

Sharks, barracudas and moray eels are notorious for their bites. Bluefish, wolffish and monkfish are equipped with formidable sets of teeth that can cause serious injuries. Walleyes and northern pike also have teeth to be avoided. Anglers should be especially cautious when removing hooks from such species. To be on the safe side, stun biting and spined fish before unhooking them.

## Bacterial Skin Infections

Some species of fish such as cunners, ocean perch, sculpins, scup and sea robins have sharp fins or rays that, though not poisonous, can lead to secondary bacterial skin infections. "Fish rose" is an infectious skin disease caused by a bacterium that lives in the surface slime of fish. This skin disease is an occupational hazard among fish handlers in processing plants. Surface cuts or abrasions on the hands of workers become infected, causing localized swelling and inflammation.

An irritation of the skin called "scombroid dermatitis" can develop from handling scombroid species (tuna, bonito, mackerel), particularly if the fish are spoiled. It is not known whether the causative agent is related to the one that produces scombroid poisoning.

*Chapter Eight*

# Fish and Nutrition

In the late 1970s, the Senate Select Committee on Nutrition and Health Needs urged Americans to replace some of the eggs, meat and dairy products in their diets with more fish, poultry and vegetable oils. This recommendation was based on the health benefits of foods that are lower in total fat, saturated fat and cholesterol—three substances that Americans consume in excess. The recommendations were confirmed in the 1980 revision of the Dietary Guidelines for Americans, a government publication with sound advice about how we should be eating.

There are at least three good nutritional reasons for including fish in our diet.
1. Most varieties of fish are high in protein and low in fat.
2. Fish are good, often excellent, sources of vitamins and minerals.
3. Fish provide particular kinds of polyunsaturated fats that are likely to reduce the risk of coronary disease and other serious illnesses. These are called omega-3 fatty acids and will be discussed in more detail later in this chapter.

In addition, there are many hundreds of different species of fish, offering an enormous range of tastes and textures. The recreational fisherman is particularly fortunate in not being restricted to the small selection generally offered by retail markets, but can eat and enjoy as many species as will take his hook.

## Nutrient Sources

Nutritionists make a good case for consuming milk products because of their calcium and riboflavin contents and for eating meat because of the protein, iron and zinc it contains. Fish (and shellfish) are equivalent to the other animal foods as a source of protein. Seafoods can also contribute significantly to our daily needs for essential minerals. Nutritionally, fish provide high protein, low fat, low saturated fat, low cholesterol (with very few exceptions), many vitamins and minerals and, uniquely, omega-3 fatty acids, which appear to have remarkable properties for preventing heart disease and inhibiting many other serious ailments.

## Proteins and Fats

Fish is high in protein and generally low in fat content. Table 7 compares protein, fat and calories in a number of protein foods.

### Polyunsaturated Fatty Acids and Heart Disease

There has been a great deal of research into the causes of coronary heart disease in recent years. We now know that the kind and amount of dietary fat we eat affects the risk of heart attack. Other factors like smoking, age, high blood pressure, obesity and blood lipid (fat) levels also affect the risk of heart disease. Simplistically, in coronary heart disease the flow of blood to the heart is restricted because of cholesterol deposits and other substances in the form of "plaque" on the walls of the arteries. A heart

Table 7: Calories, Fat and Protein in 100 Grams (3½ Ounces) of Selected Raw Protein Sources

| Raw Protein Source | Protein | Fat | Calories |
|---|---|---|---|
| Hamburger, regular ground[1] | 20.7 | 10.0 | 179 |
| American cheese[1] | 25.0 | 32.2 | 398 |
| Chicken, dark meat, without skin[1] | 18.1 | 3.8 | 112 |
| Eggs, whole, fresh[1] | 12.9 | 11.5 | 163 |
| Lamb, leg[1] | 17.8 | 16.2 | 222 |
| Peanuts, shelled[1] | 26.0 | 47.5 | 564 |
| Peanut butter[1] | 25.2 | 50.6 | 589 |
| Ham[1] | 15.9 | 26.6 | 308 |
| Bluefish[2] | 20.0 | 4.2 | 124 |
| Haddock[2] | 18.9 | 0.7 | 87 |
| Mackerel[2] | 18.6 | 13.9 | 205 |
| Bluefin tuna[2] | 23.3 | 4.9 | 144 |
| Northern pike[2] | 19.3 | 0.7 | 88 |
| Flatfish[2] | 18.9 | 1.2 | 91 |

[1] Source: Watt, B.K., and Merrill, A.L., 1963, *Composition of Foods*, Agriculture Handbook No. 8, USDA, Washington, D.C.

[2] Source: Exler, J., 1987, *Composition of Foods: Finfish and Shellfish Products*, Handbook No. 8-15, USDA, Washington, D.C.

attack occurs when a blood vessel to the heart becomes completely obstructed and blood flow is cut off.

Ways of discouraging the build-up of plaque also discourage the progress of heart disease. Keeping the intake of fat down and limiting saturated (mostly animal) fat and cholesterol helps lower blood lipid levels. Eating high-fiber foods also helps, though it is not clear why or how. People in high-risk categories do well to restrict their intake of foods rich in cholesterol and saturated fats.

Many feeding studies on both animals and humans show that the replacement of meat with fish significantly lowers blood cholesterol levels. Heart attacks are virtually unknown among Eskimos, whose chief food supply consists of fish and marine animals.

## Omega-3 Fatty Acids

It appears that a constituent of the fat in fish—omega-3 fatty acids—helps to prevent the build-up of plaque and so helps to reduce the risk of heart disease. These particularly important omega-3s are found only in fish and shellfish. None are found on land or in land animals. Fatty fish have more omega-3s than lean fish, but all fish have some.

Table 8: Omega-3 Fatty Acid Content of Various Fish

gm/100 gm raw edible portion (percentage content in raw fish)

|  | |
|---|---|
| Bass (freshwater, mixed species) | .6 |
| Bass, striped | .8 |
| Bluefish | .8 |
| Burbot | .2 |
| Carp | .4 |
| Catfish, channel | .4 |
| Cisco | .4 |
| Cod, Atlantic | .2 |
| Cod, Pacific | .2 |
| Croaker, Atlantic | .2 |
| Drum (freshwater) | .5 |
| Flatfish (mixed species) | .2 |
| Grouper (mixed species) | .2 |
| Haddock | .2 |
| Lingcod | .2 |
| Mackerel, Atlantic | 2.3 |
| Mackerel, king | .3 |
| Mackerel, Pacific and jack | 1.4 |

(Cont'd)

Table 8: Omega-3 Fatty Acid Content of Various Fish (Continued)

| Fish | Omega-3 |
|---|---|
| Mackerel, Spanish | 1.3 |
| Mahi-mahi | .1 |
| Mullet, striped | .3 |
| Perch (mixed species) | .3 |
| Pike, northern | .1 |
| Pike, walleye | .3 |
| Pollock, Atlantic | .4 |
| Pollock, walleye | .4 |
| Pompano (Florida) | .6 |
| Rockfish, Pacific | .3 |
| Roe (mixed species) | 2.4 |
| Sablefish | 1.4 |
| Salmon, Atlantic | 1.4 |
| Salmon, chinook | 1.4 |
| Salmon, chum | .6 |
| Salmon, coho | .8 |
| Salmon, pink | 1.0 |
| Salmon, sockeye | 1.2 |
| Sea Bass (mixed species) | .6 |
| Sea Trout | .4 |
| Shark (mixed species) | .8 |
| Sheepshead | .3 |
| Snapper (mixed species) | .3 |
| Squid (mixed species) | .5 |
| Sucker | .5 |
| Sunfish, pumpkinseed | .1 |
| Swordfish | .6 |
| Tilefish | .4 |
| Trout (mixed species) | .7 |
| Trout, rainbow | .6 |
| Tuna, bluefin | 1.2 |
| Tuna, skipjack | .3 |
| Tuna, yellowfin | .2 |
| Whitefish (mixed species) | 1.3 |
| Whiting (mixed species) | .2 |
| Wolffish, Atlantic | .6 |

Source: Exler, J., 1987, *Composition of Foods: Finfish and Shellfish Products*, Handbook No. 8-15, USDA, Washington, D.C.

Omega-3s are now credited for many other disease-preventing functions. Arthritis, multiple sclerosis, bronchial asthma, diabetes, migraines, even breast cancer may all be reduced in people eating small quantities of omega-3s from fish.

Omega-3s (called n-3s by some scientists) are a type of polyunsaturated fat, but it appears that only this omega-3 type provides these special health benefits. All polyunsaturated fats are generally better for you than saturated fats.

## Cholesterol

Cholesterol levels in fish are very low. High levels of cholesterol are associated with heart disease. Consumption of foods with high levels of cholesterol can add significantly to the levels in the blood of some people. Seafood not only has omega-3s, which help to reduce the risks of heart attacks, it also contains less cholesterol, which is a substance heavily implicated as a cause of heart disease.

All finfish are low in cholesterol. Crabs, lobsters and shrimp have moderate levels. Some squid, however, is quite high. But even these contain the omega-3 fats, which help to counteract the bad effects of the cholesterol.

Table 9 shows cholesterol levels in some fish. Levels under 100 mg per 100 gm of edible fish are all considered to be low. To compare, dark chicken meat without skin has about 360 mg/100 gm and American cheese 450 mg/100 gm.

Note that both fat and cholesterol levels of fish increase enormously if you fry the fish in saturated fats. Breading particularly picks up a great deal of fat. In order to retain the nutritional advantages of fish, it is necessary to cook it in ways that do not add other undesirable things. Avoid deep-fat frying and do not use heavy, rich sauces with fish. One of the healthiest ways to prepare fish is to broil it—a particularly easy method for many anglers.

Fat and, to some extent, protein in fish depend on what they eat, as well as on age, sex, spawning cycle, season and geography. Fish caught during the spawning season or in waters having sparse food supplies have a lower fat content than usual.

Table 9: Cholesterol Content of Selected Raw Protein Sources

| Food | Cholesterol mg/100gm |
|---|---|
| American cheese[1] | 450 |
| Chicken fryers[1] | 360 |
| Eggs, whole, fresh[1] | 2000 |
| Lamb, leg[1] | 320 |
| Ham[1] | 300 |
| Bluefish[2] | 59 |
| Cod, Atlantic[2] | 43 |
| Butterfish[2] | 65 |
| Monkfish[2] | 25 |
| Ocean Perch[2] | 54 |

[1] Source: Watt, B.K., and Merrill, A.L., 1963, *Composition of Foods*, Agriculture Handbook No. 8., USDA, Washington, D. C.

[2] Source: Exler, J., 1987, *Composition of Foods: Finfish and Shellfish Products*, Handbook No. 8-15, USDA, Washington, D.C.

Variations also occur in the distribution of the fat within each fish. The fattier portions of a fish are the nape, the belly flap and the dark muscle. Dieters who want to cut back on calories should avoid these sections, while others who are not dieting may enjoy them to receive the full benefits of the omega-3s. As well as being higher in fat, the dark meat is also higher in desirable minerals.

## Vitamins and Minerals

Fish are good sources of potassium and phosphorus and saltwater fish are an almost unique source of fluorine and iodine. Seafoods are among the richest food sources of iodine, which is essential for the formation of hormones that regulate growth and the rate of metabolism.

Fish is low in sodium, a fact that often surprises people confined to low-sodium diets, many of whom are wrongly avoiding saltwater fish. Low-sodium diets recommend less than 100 milligrams (preferably less than 60 milligrams) of sodium per serving. Most saltwater and freshwater fish in 100 gram portions fit very nicely into low-sodium diets (note that 100 grams equals 3½ ounces and 1 milligram equals $1/1000$ gram). Anglers have an advantage over other fish eaters in that they know their fish is not processed or preserved with sodium additives, as is sometimes done commercially.

Fish are good sources of the water-soluble B vitamins. The body has little capacity for storing vitamins that dissolve in water, so it needs a new supply daily. Thiamine ($B_1$), riboflavin ($B_2$), pantothenic acid ($B_3$) and niacin are essential for release of energy from carbohydrates, fats and proteins. These and other B vitamins are present in useful amounts in most fish.

Vitamin $B_{12}$, which is necessary for growth and blood formation, is found only in meat, poultry, fish and shellfish.

B vitamins concentrate in the dark muscles of fish at levels as much as ten times higher than in the light meat. Baking is thought to retain more of the B vitamins than does frying.

Vitamins A, D and E are fat-soluble. Consequently, they are readily stored in the body for use when intake of them is low. Fish liver oils are very rich sources of these vitamins, but are seldom consumed. Because fat-soluble vitamins are retained in the body, it is important not to ingest too much in one dose. For example, levels of vitamins A and D in fish liver oil may be so great as to be hazardous to health. Because of this, some fish oil preparations actually remove much of the vitamins. Further, because the liver is where organic substances are stored and detoxified, consuming too much fish liver oil may carry the risk of absorbing dangerous levels of contaminants. It is far better to get the health benefits of fish from eating fish and not from taking fish oil supplements.

Experts studying fish oils and their role in health and nutrition are recommending that people eat fish at least three times a week and can demonstrate significant health benefits from following that advice.

Eating everything in moderation assures having a balanced diet. Anglers who throw some of their catches back into the sea should retain and enjoy them next time, knowing that a new and wide variety of nutrients are added to their food as a result.

Chapter Nine

# Basic Techniques for Cooking Fish

You can cook nearly any fish almost any way with excellent results. Broiling species rich in fat is healthier because it allows some of the fat to run off. Leaner fish is generally better steamed. Although the flavor, texture, appearance and size of fish vary enormously, the fundamental rules for cooking are few and easy to follow.

## The Canadian Cooking Rule

This is the simplest way to cook fish. It works for all finfish—not for shellfish. This method was extensively tested by Canadian government agencies and works every time. The rule is "10 minutes to the inch" for cooking fish.

Let's expand that a little.

To bake, preheat the oven to an accurate 450°F. Measure the thickness of the fish at its thickest point, as the fish sits on its cooking container.

For every inch of thickness, cook fresh fish for 10 minutes. This means that if it is half an inch thick, cook it for 5 minutes; if it is 2 inches thick, cook it for 20 minutes. Always set the oven at 450°F.

Pieces an inch thick or more cook more evenly if you flip them over about two-thirds of the way through the cooking time.

If the fish is frozen, double the cooking time to 20 minutes for each inch at 450°F.

The rule applies equally well to every other cooking method.

Broil under a preheated broiler with the fish 2 to 4 inches from the heat for 10 minutes for each inch, or 20 minutes for each inch of frozen fish.

When poaching, boiling or steaming, time from the point the water returns to a boil. Cook for 10 minutes per inch, or 20 minutes per inch for frozen fish.

Fish is often overcooked. This toughens it, dries it out and destroys the fine flavor. Learning to cook fish just enough is easy. When the flesh flakes easily from the bones, the fish is moist and tender and has its best, delicate flavor.

Raw fish has a watery, translucent appearance. Cooking causes the juices to turn milky. The flesh becomes opaque, with a white tint. Fish is completely cooked when the flesh is opaque in the center of the thickest part.

One frequent problem is that fish fillets are not evenly thick. The tail end is much thinner than the head end, so that the tail overcooks before the thicker part is done. There are two easy ways to resolve this. Either fold the tail end underneath the fillet, or, when cooking several fillets at once, overlap the tail ends. Both methods give a more uniform thickness that cooks evenly.

## Baking

Place the fish on a lightly greased pan. If the skin is left on the fillet to hold it together and you do not want the skin on the plate, put the fillet, skin side down, on ungreased foil on a baking pan. When the fish is cooked, it slides off the skin easily, leaving the skin and foil stuck together for easy disposal. You can also put the fish on a layer of vegetables in the bottom of a baking dish. This makes the fish especially moist and avoids having to add any fat.

It is not necessary to baste when using the Canadian cooking rule. The top does not dry out unless the fish is very thick. If it is more than 2 inches thick, baste during the second half of cooking with pan juices or a small amount of olive oil, tomato sauce, salad dressing or a marinade. Turning the fish over after two-thirds of the cooking time also serves to avoid any need to baste.

One method used to prevent drying out during oven baking is to sprinkle about one ounce of water onto the baking pan around the fish.

Fat fish and lean fish both bake well. Bluefish, lake trout, salmon, steelhead, tuna and other fatty fish need light seasoning only, but their robust flavors can also stand the addition of strong accompaniments. Large lean fish such as burbot, mahi-mahi, striped bass, walleye and cod may need a small amount of sauce or oil for basting and seasoning.

Proper baking produces tender, moist fish and is one of the easiest cooking methods possible.

## Oven-frying

Oven-frying is a simple technique that combines the ease of baking with the crispness of pan-frying (see below), but without the addition of large amounts of fat. Cut fish into serving-size pieces, dip in seasoned milk, coat with toasted, fine, dry crumbs and place in a shallow, well-greased baking pan. Sprinkle one tablespoon of vegetable oil over the fish, then bake once again using the Canadian Cooking Rule: at 450°F for 10 minutes to the inch (or 20 minutes per inch for frozen fish). The fish does not need turning, basting or surveillance.

## Broiling

Like baking, broiling is a dry-heat cooking method. However, broiling is more likely to dry out the surface of the fish because the heat is closer and more intense. Broiled fish normally needs turning over, unless it is under one half an inch thick where the heat of the pan cooks the underside.

When broiling, it is generally necessary to baste the fish with a light sauce to help keep the fish moist.

## Charcoal Broiling

Fish is a natural for dry-heat cooking over hot coals because it cooks quickly. Pan-dressed fish, fillets and steaks are all suitable for charcoal broiling or gas-grilling. Because fish flakes and tends to crumble as it cooks, a long-handled, hinged, wire grill (coat it with a nonstick spray before use) that holds the fish is a useful implement. So is a grill insert with small holes. Take particular care when turning and serving the fish. Basting with almost any barbecue sauce before and during cooking helps to retain moisture and add flavor.

Covered barbecue grills are even better for cooking fish because they steam the fish rather than char it. Put the fish on a lightly greased foil pan (skin side down for fillets) and cook the fish with the grill cover closed just as you would in the oven. The smoke from the barbecue flavors the fish, which needs no turning or basting, though you may like to use a sauce to accentuate the barbecue flavors. Before you start to cook, separate the hot coals into two heaps, one on each side of the grill. Put the fish in the middle of the grill so that it is not directly over the heat source. Unless your coals are really hot, it is unlikely that your grill will reach 450°F, so cooking time will be a little longer than suggested by the Canadian

Cooking Rule. For this one, test the fish with a fork to see if it is done.

## Frying

Frying has the advantages of cooking quickly and retaining the product's moisture and nutrients. Its disadvantages are that it adds large amounts of fat we do not need and can be a safety hazard in the kitchen. Frying is not recommended for those watching their weight or for anyone seeking the most healthful ways of cooking fish.

Choose a fat that heats to a high temperature without smoking. Smoking fat begins to decompose and gives an unpleasant flavor to the fish. It also breaks down and becomes a possible health hazard, as does fat that is reused too often. Therefore, it is important to be careful about reusing fat, whether for pan-frying or deep-fat frying. Vegetable oils are better than animal fats because they begin smoking at higher temperatures. Temperature of the fryer fat is extremely important: too high a heat browns the outside before the center cooks; too low a heat yields a pale, greasy, fat-soaked product. A temperature of $365°F$ is best for frying fish. Drain fish on absorbent paper after frying to remove excess fat. Keep the fish that are fried first warm in a low-temperature oven until you fry all the remaining pieces. Lean to moderately fat fish are best for frying. Fish that are high in fat yield too "rich" a product.

## Deep-fat Frying

Use enough fat to float the fish but do not fill the fryer more than half full. Allow enough room for the fish in the bubbling fat.

Dip the fish in a liquid, either milk or water or milk and eggs, and coat with a breading, or alternatively dip the fish in a batter. The coating keeps the fish moist and holds the fish together, yet it also increases the amount of fat it absorbs.

Adding baking powder to the batter mixture produces a "puff" batter that, when cooked, puffs out and is somewhat moist inside. A good recipe for puff batter:

*Ingredients*

- 2 cups of sifted flour
- 1 teaspoon of baking powder
- ¼ teaspoon of salt
- 2 eggs
- 1 cup of milk or water

Beat the eggs, add the milk or water to the eggs and pour into the dry ingredients. Beat the mixture until smooth. Use beer or dark beer in place of the milk or water for a slightly different flavor.

Put only one layer of fish in the fryer at a time, allowing enough room to keep the pieces from touching. This prevents the temperature from dropping too far and assures thorough cooking and even browning. The fat is at the right temperature if a crust forms almost immediately after the fish pieces are added. Fry the fish until they are golden brown and the pieces float. About three to five minutes is usually long enough.

# Pan-frying (Sautéing)

The basic method for pan-frying is to dip the fish in a liquid, coat with a breading and then fry in a small amount of fat. Within this method there are numerous variations affecting the flavor and texture of the fish, as well as its nutritional value (because of the fat and the type of fat that you add during cooking).

Dip small pan-dressed fish or fillets in water or milk, then in flour or cornmeal and fry. Or, you can use beaten eggs thinned slightly with water or milk for the liquid and breading mix for the coating. There are hundreds of possible coatings, commercial and homemade. In the end it depends on what you like best.

Use either margarine or vegetable oil, or a combination of butter and oil. Remember that butter adds saturated fat and cholesterol, two substances that many of us are trying to avoid.

Cook one layer of coated fish at a time, taking care not to overload the pan and cool the fat. Turn the fish over only once—when the first side is brown.

## Stir-frying

This is a quick and easy method. Use a large frying pan or skillet or, of course, a wok. Add 1 tablespoon of vegetable oil to the pan and heat it to about $375°F$.

Cut fish into one inch cubes and marinate to firm and flavor it.

Add the marinated fish to the pan and stir vigorously so that it is turning constantly.

Do not add more than 8 ounces of fish to the pan at a time. This amount cooks in about 2-3 minutes. Then add more oil and another batch of fish. Too much fish reduces the pan's heat and the fish does not cook fast enough.

Cut vegetables into strips and stir-fry before adding the fish. Ginger and garlic in generous quantities make the fish taste authentically "Chinese."

Stir-frying adds very little fat and it retains nutrients well because it cooks so quickly. It is also very easy and you can adapt it for outdoor cooking.

## Poaching and Steaming

The two moist-heat cooking methods, poaching and steaming, are closely related, the difference being in the amount of cooking liquid used. Fish cooked in moist heat requires very little cooking time and is usually served with a sauce, often made from the cooking liquid itself. Firm-fleshed fish such as salmon or trout or low-fat fish like sole or cod, whole or cut up, are the best choices for poaching or steaming. Very fat fish such as mackerel and herring or soft-textured fish like whiting and shad tend to fall apart when cooked in liquid.

### Poaching

Cooking in a simmering liquid is called poaching. Place the cleaned fish in a wide, shallow pan (such as a large fry pan) and barely cover it with liquid. The liquids you use may be lightly salted water, water seasoned with spices and herbs, milk, or a mixture of white wine and water. You can also use court bouillon or fish stock. In a covered pan, simmer the fish in the liquid just until the fish flakes easily. Plan the cooking time according to the Canadian Cooking Rule. Reduce and thicken the liquid — which contains flavorful juices — to make a sauce for the fish. Serve poached fish with a sauce, or use it as the main ingredient in a casserole or other combination dish. Poached fish is delicious in salads when chilled and flaked.

### Steaming

Cooking in steam generated from boiling water is another method for preparing your fish. Fish retain their natural juices and flavors when cooked over moisture in a tightly covered, deep pan. Use a steaming rack, vegetable steamer or any device that prevents the fish from touching the water. The water may be plain or seasoned with various spices, herbs or wine. Bring the water rapidly to a boil, place the fish on the rack, cover the pan tightly and steam for the time determined by the Canadian Cooking Rule. Serve the steamed fish the same way you serve poached fish. You can steam even the most delicate fish without it falling apart.

## Braising and Stewing

Braising and stewing are based on poaching. Stewing uses plenty of liquid—at least enough to cover the ingredients. Braising usually requires only a small amount of liquid, not enough to cover the food. Both techniques use liquid as the cooking medium and always include vegetables and herbs. Serve the cooking liquid and its aromatic ingredients as part of the finished dish. Neither stewing nor braising of fish is a long cooking process. Simmering fish for hours causes the fish to disintegrate. Use relatively firm-fleshed fish (for example, blackbass, burbot, carp, catfish, cod, eel, haddock, rockfishes, sea bass, pollock, sauger, walleye and cusk) to get the best flavor and texture from either of these cooking methods. Very fatty fish such as mackerel or soft fish such as weakfish tend to fall apart when cooked in liquid. Thin fillets are too delicate to use this way, whereas thick ones, cut into chunks, are excellent.

### Braising

The process of getting a concentrated exchange of flavors among various ingredients is called braising. Strong-flavored oily fish are good for braising. Vegetables are used primarily as flavoring agents, and usually only a single kind of fish is braised at a time. Use a minimum amount of liquid in order to get as concentrated a blend as possible. Because you cannot simmer the fish for a long time, sauté the vegetables first and then cook them in the broth to release their flavors before adding the fish. Simmer briefly. After cooking, combine the vegetables (often puréed) and the liquid with other ingredients to make a sauce for the fish.

## Stewing

Most fish stews contain several kinds of fish and vegetables and are simmered in a lot of liquid. Cook the vegetables first, in a broth, almost to completion. Many traditional recipes use a fish stock (after discarding the fish and shellfish) for cooking the vegetables. Add the fish to the broth and cook for a few minutes only until it flakes. This procedure assures retention of the textures and flavors of the individual ingredients, while giving the flavors a chance to blend. Serve the stew straight from the pot with no further preparation.

## Microwave Cooking

Microwave ovens can be one of the best ways of cooking fish, especially thinner fillets. Since ovens vary so much in power and cooking speed, you have to experiment a little with your own oven to get the best results. Cook for a shorter time than you expect the fish to need, then cook a few seconds more if necessary until the fish is done. Bear in mind that half a minute is a long time in a microwave. Cook little and test often is the rule. Also remember that food continues to cook for a short time after it is removed from the microwave, so fish that is just turning opaque will continue to firm up for a while.

Fresh fish needs no seasoning in the microwave. Simply cook it as it is. Keep the thickness of the fish as even as possible: overlap or fold fillets if necessary.

Microwaving thicker steaks or whole fish is more difficult. Unless you have considerable experience with the microwave oven, these are generally much more successfully cooked (according to the Canadian Rule) in the regular oven, as described earlier.

## Wines in Fish Cookery

Fish dishes often include dry wine as an ingredient. When poaching or steaming fish, use a court bouillon or a fish fumet as the cooking liquid. Make a court bouillon by simmering water and wine or vinegar with vegetables and herbs for flavoring. Add wine at the beginning of the simmering process to produce a mild-tasting bouillon; for a more pronounced wine flavor, add the wine halfway through.

A fish fumet is a fish stock made from water, wine and fish or fish trimmings. Again, you may flavor it with vegetables and herbs. Fumets should be more pronounced and concentrated in flavor than bouillons, so use less water. After poaching and straining the fumet, use it as the base for a sauce, a stew or for braising. Store it in the freezer.

You may also use wines directly in stews, braises or sauces.

## Using Leftover Fish

You do not have to throw away fish left over at the end of a meal. Flake leftover poached, steamed or baked fish and use it as the main ingredient of an excellent cold salad. Mix it with a little mayonnaise, onions, celery and some spices and enjoy it as is or in a sandwich. You can even mix cooked fish with a white sauce for a macaroni, rice, or vegetable-based casserole.

## Cookbooks

There are hundreds, perhaps thousands, of fish cookbooks. A very small number of these are mentioned in the "Further Reading" section below. Fish recipes can be extremely simple—like using the Canadian Cooking Rule and baking your catch—or very complex, with sauces and lots of ingredients.

Whichever way your culinary preference runs, the end product you put on the table will be much better if the fish you use has been properly handled and cared for. Follow the advice in this book and be sure of enjoying your catch.

# Glossary

**Aerobic:** Living or active with air.
**Allergen:** Agent found in some foods that stimulates the production of specific antibodies creating sensitivity to that food. Subsequent meals of the same food will induce an allergic reaction and sometimes anaphylactic shock. Symptoms include hives, headaches, swelling and gastrointestinal disturbances.
**Anadromous:** Sea fish that migrate to freshwater rivers, streams and lakes to spawn.
**Anaerobic:** Living or active without air.
**Anchorworms:** Parasitic copepods that have anchorlike projections growing out of their heads into the flesh of the fish (host).
**Anisakiasis:** Infection caused by eating fish infested with the roundworm *Anisakis simplex*. Most cases involve humans who have eaten raw or undercooked fish or lightly cured or slightly salted herring. The anisakis larvae do not mature and reproduce in the human gut but can cause inflammation and ulceration.
**Anisakis simplex:** Roundworm that infests herring and mackerel in the northeast United States. Eating fish infested with this parasite can cause the infection "anisakiasis."
**Antigen:** When introduced into the body, this substance stimulates the production of an antibody.

**Bacalao:** Salt cod. Headed, gutted fish with the larger part of the backbone removed. Fish is cleaned and washed, then preserved with dry salt for three weeks. Salted fish is washed, salted again and dried.

**Basting:** Moistening with liquids periodically while cooking.

**Bleeding:** Bleeding your fish lightens the flesh color and helps it keep longer. This applies especially to oily fish.

**Botulism:** Disease caused by *Clostridium botulinum* (see *Clostridium botulinum*).

**Braising:** A method of cooking using a small amount of liquid to get a concentrated exchange of flavors among various ingredients.

**Brevitoxin:** Toxin, produced by "red tide," that causes paralytic shellfish poisoning. The poison acts on the human nervous system and is one of the most potent toxins known to humans. There is no known antidote. (see Red tide).

**Brine:** Saltwater solution.

**Burlap:** Coarse fabric of hemp or jute, usually made into bags.

**Buttons:** Anchorworms that have an anchorlike projection of the head growing into the flesh of the fish, especially on ocean perch, bluegills and largemouth bass.

**Canadian Cooking Rule:** This method, which was tested extensively by Canadian government agencies, is one of the simplest and best ways to cook finfish. The rule is to cook the fish 10 minutes to the inch (see Chapter Nine).

**Candling:** Common commercial practice for locating and eliminating parasites. Fillet is placed on a glass surface with a strong light underneath. The light produces a silhouette of the parasites, which are then removed.

**Caviar:** Salted sturgeon roe.

**Carnivorous:** Feeding on animal tissue.

**Cestodes:** Parasites (tapeworms) that live in intestines.

**Chowder:** Thick, stew-like soup usually based on clams, but may contain other fish and shellfish.

**Ciguatera poisoning:** Toxin found mostly in species from tropical and subtropical areas. Toxin is thought to be from one of the dinoflagellates consumed by fish indirectly through the food chain. Larger fish are more suspect. Symptoms include gastrointestinal disturbances (vomiting, cramps and diarrhea), aching muscles and joints, extreme weakness, tingling sensations, inverse temperature sensitivity and intense itching. Symptoms occur within six hours. Toxin is also heat resistant.

**Clostridium botulinum:** Causes the disease botulism. Symptoms include muscular and respiratory paralysis. Can only grow in anaerobic conditions.

**Copepods:** External parasites.

**Curing, pickle:** Wet salting. Fish are salted in a container so that they are cured or pickled in the liquid that forms.

**Dehydration:** Loss of moisture. It is important to wrap product well to prevent dehydration during frozen storage.

**Diphyllobothriasis:** Infection caused by eating fish infested with the parasite *Diphyllobothrium latum* (fish tapeworm). Symptoms of this infection include abdominal discomfort, diarrhea or constipation.

**Diphyllobothrium latum:** Fish tapeworm found in some freshwater fish such as northern pike and yellow perch. Eating an infected fish can cause diphyllobothriasis.

**Dorsal:** Relating to the back.

**Dressing:** Removing the viscera from the fish.

**Drip loss:** Fish loses moisture and soluble proteins when thawing. Some loss of weight occurs naturally in fresh, unfrozen fish.

**Dysentery:** Disease distinguished by symptoms of diarrhea with blood and mucus in the feces.

**Enzymes:** Complex, mostly protein products of living cells that induce or speed up a chemical reaction in plants or animals without the enzymes being permanently altered.

**Eviscerating:** Removing the viscera from the fish.

**Fats, polyunsaturated:** Fats that can accept four or more hydrogen atoms in the molecule. These are usually liquid at room temperature.

**Fats, saturated:** Fats that cannot take more hydrogen atoms into their molecules. Normally solid at room temperature.

**Fat-soluble vitamins:** Vitamins that dissolve in fat and therefore are present in body fats. Vitamins A, D, E and K are fat-soluble vitamins.

**Fillet knife:** Knife with a thin blade used to fillet fish; sharp, but not sharp enough to cut through the bone.

**Fillets:** "Slices of practically boneless fish flesh of irregular size and shape, which are removed from the carcass by cuts made parallel to the backbone and sections of such fillets cut so as to facilitate packing," U.S. Grade Standards, Title 50, U.S. Code.

**Finnan haddie:** Smoked haddock (see Chapter Five).

**Fish-and-chips:** Battered and fried white-meat fish and french fried potatoes.

**Fish lice:** External parasites.

**Fish rose:** Infectious skin disease caused by a bacterium living in the surface slime of fish. Surface cuts or abrasions on fish handlers' hands become infected, causing localized swelling and inflammation.

**Flatworms:** Trematodes; parasites.

**Freezer burn:** Dehydration of a frozen product caused by loss of moisture during evaporation. It is recognized by a whitish-yellowish, cottony appearance of the flesh, especially at the cut edges or thinner places. Can encourage rancidity.

**Frill:** Bony fringe surrounding the body of the flatfish; it is part of the fin structure.

**Fugu poisoning:** Tetrodotoxin poisoning caused by eating tainted puffer fish. Symptoms include tingling of the lips, mouth and extremities, followed by numbness of the limbs and general loss of muscular coordination. Symptoms appear soon after eating, and death may follow within two to twelve hours. Mortality rate may be as high as 50 to 60 percent.

**Fumet:** Fish stock.

**Gaff:** A metal hook for holding or lifting heavy fish.

**Gills:** Parts of the fish that are used for obtaining oxygen from water.

**Glaze:** Protective coating of ice on frozen product to prevent dehydration in the fish.

**Gravlax:** Also called gravadlax, dill salmon or marinated salmon. Gravlax is sides (fillets) of salmon dry-marinated in a mixture of salt, sugar and dill weed. It is similar to smoked salmon in taste and usage.

**Green-salted:** Fish that have been stacked and salted for two days (see Kenching).

**Grubs, black:** Trematodes found on the skin of freshwater fish such as yellow perch, sunfish and bass. They are usually found near the tail. The affected fish is edible after you remove the grubs with a knife.

**Grubs, yellow:** Trematodes found on the skin of freshwater fish such as yellow perch, sunfish and bass. They are usually found near the tail. The affected fish is edible after you remove the grubs with a knife.

**Gutting:** Removing the viscera from the fish.

**Heavy metals:** Found in virtually all plants and animals. Heavy metals include lead, cadmium and mercury, all of which are toxic at high levels. Heavy-metal poisoning in humans can result in neurological disturbances, congenital birth defects and even death.

**Hepatitis:** Viral disease that can be caused by eating shellfish harvested from polluted waters. Hepatitis is characterized by the inflammation of the liver.

**Histamines:** Chemicals produced by decomposition in some fatty species.

**Hydrometer:** A floating instrument for determining specific gravities of liquids and their strength.

**Kenching:** Dry salting. The process of stacking fish with a layer of salt separating each layer of fish. Kenching extracts most of the moisture from the fish.

**Kidney:** Dark red organ along the spine of a fish inside the belly cavity. Remove the fish's kidney by scraping with a knife handle or a stiff brush.

**Marinating:** Soaking in a brine or acidic solution such as vinegar.

**Meltwater:** Water from melted ice.

**Membrane:** A thin layer of plant or animal tissue.

**Methylmercury:** A highly toxic form of metallic mercury produced by bacteria in sediments of the ocean bottom.

**Microsporidia:** Single-celled internal parasites.

**Milts:** Male gonads of fish.

**Molluscs:** Shellfish such as clams, mussels, oysters, conch, snails and abalone. Squid and octopus are also molluscs.

**Mouse unit:** A measure of a toxin's deadliness. Extract of fish flesh is injected into mice to determine a lethal dose, which is the amount required to kill a 20-gram mouse in 15 minutes.

**M.U.:** Mouse unit. See above.

**Myxosporidia:** Single-celled internal parasites.

**Myxosporidian parasites:** Microscopic parasites that form black clusters on fish, giving a peppered appearance to the flesh. These parasites also secrete a protein-digesting enzyme that turns fish flesh mushy or jellylike. They can also produce small white "pus pockets" on certain fish found along the eastern United States coast.

**Nape:** Front part of fillet that encloses the gut cavity of the fish.

**Necks:** Clam's siphon that protrudes beyond its shell.

**Nematodes:** Long cylindrical parasitic worms.

**Nobbing:** Removing the head and viscera from a fish without cutting the belly. This leaves the roe or milt in the fish. Nobbing is used mainly on herring.

**Omega-3s:** Polyunsaturated fats found in fish and shellfish. Omega-3s are credited for reducing heart disease, arthritis, multiple sclerosis, bronchial asthma, diabetes, migraines and breast cancer.

**Pan-dressed or pan-ready:** Fish with head, viscera, fins and tail removed.

**Paralytic shellfish poisoning:** Caused by "red tide" or plankton bloom. Eating shellfish containing saxitoxin (a toxin produced by poisonous dinoflagellates) causes PSP. Symptoms appear immediately after consuming contaminated shellfish. Symptoms are tingling of the lips, mouth and extremities, followed by numbness in the arms and legs and loss of muscular coordination. Breathing becomes difficult as the illness progresses and death within two to twelve hours may result from paralysis.

**Parasites:** A plant or animal living in or on a host. Surface or external parasites include leeches, fish lice and copepods. Internal parasites include tapeworms, flatworms and roundworms and single-celled organisms.

**PCBs:** Polychlorinated biphenyls. Used primarily as a coolant insulator in electrical transformers and capacitors. Has led to the contamination of major inland waterways in the United States. It is toxic to man and animals, perhaps even cancer-causing.

**Pectoral:** Relating to the breast or chest.

**Pellicle:** Shiny, hard outer crust that forms on the outside of drying fish during the smoking process.

**Pen:** Chitinous "bone" inside a squid.

**Phocanema decipiens:** Roundworm belonging to the same family as Anisakis. Cod flesh is sometimes infested with them, as may be smelt, haddock, plaice, ocean perch and whiting. Known as cod worms, they are mostly found on larger specimens and also in fish that inhabit shore areas where seals reside. Cod worms may cause gastric disturbances in humans.

**Poaching:** Cooking in a simmering liquid.

**Preservation:** Processing food to keep it from spoiling.

**Protozoans:** Single-celled internal parasites.

**PSP:** See Paralytic shellfish poisoning.

**Puffer-fish poisoning:** Tetrodotoxin poisoning. Symptoms include tingling of the lips, mouth and extremities, followed by numbness of the limbs and general loss of muscular coordination. Symptoms appear soon after eating the poison and death may follow within two to twelve hours. Mortality rate may be as high as 50 to 60 percent.

**Puffers:** Blowfish. Many puffers from Pacific waters have internal organs that are highly toxic. Puffers from Atlantic waters are not as toxic.
**Rancidity:** The oxidation of the natural oil in the fish, making the fish unpalatable.
**Red tide:** A plankton bloom. See Paralytic shellfish poisoning.
**Retardant:** A substance (example: lemon juice) that slows down spoilage of fish.
**Rigor mortis:** Process of a body stiffening after death.
**Rock salmon:** Previous name for dogfish in England. The name is now prohibited.
**Roe:** Fish eggs.
**Roundworms:** Long cylindrical parasitic worms.
**Salinometer:** A hydrometer designed to measure the specific gravity of brine.
**Salmonella:** Naturally occurring pathogenic bacteria associated with polluted waters and unsanitary handling of fish and shellfish. Improper temperatures above 40°F allow rapid growth of this bacteria. Symptoms of salmonella poisoning are vomiting, nausea and diarrhea.
**Salt cod:** Bacalao. Headed, gutted fish with the larger part of the backbone removed. Fish is cleaned and washed, then preserved with dry salt for three weeks. Salted fish is washed, salted again and dried.
**Salt fish:** Headed, gutted with the larger part of the backbone removed. The fish is cleaned and washed, then preserved with dry salt for three weeks. Sold whole or in fillets. Keeps well in cool temperatures.
**Salt, kosher:** This does not have added iodine salts. Kosher salt is pure and is recommended for salting fish.
**Sautéing:** Frying lightly in a little fat or oil.
**Saxitoxin:** Toxin produced by red tide, which causes paralytic shellfish poisoning. Poison acts on the

human nervous system and is one of the most potent toxins known to man. There is no known antidote.

**Schillerlocken:** Smoked dogfish belly flaps.

**Scombroid dermatitis:** Skin irritation caused by handling scombroid species (tuna, bonito, mackerel), particularly when handling spoiled fish. It is not known whether the causative agent is related to the one that produces scombroid poisoning.

**Scombroid poisoning:** Fish such as tuna, bonito, mackerel-type fish, swordfish, mahi-mahi and bluefish may cause scombroid poisoning if exposed to warm temperatures for several hours.

**Scombrotoxin:** Heat-resistant toxin found in certain fish (tuna, bonito, mackerel-type fish, swordfish, mahi-mahi and bluefish) when exposed to warm temperatures after catching.

**Seawater, chilled:** Mixture of clean seawater and crushed or flaked ice.

**Selenium:** A nutrient obtained from seafoods, especially shellfish and tunas, as well as from kidneys, grains, beans and nuts. It may diminish the toxicity of ingested mercury and may also play a part in cancer prevention.

**Seviche:** Also spelled **Ceviche**. Fish "cooked" in acid, usually lemon juice, lime juice or vinegar instead of heat. Marinate for 24 hours or until fish loses its translucence and turns white.

**Shigellosis:** A type of dysentery caused by the Shigella bacteria found in estuarine waters polluted by the discharge of raw sewage or of sewage effluent. Mostly found in bivalve molluscs.

**Siphon:** Neck of the clam. Clams may carry red-tide toxin here as well as in their digestive glands.

**Spawning:** Releasing eggs; fertilizing eggs.

**Staphylococcus aureus**: Bacteria generally absent from the ocean environment and seafoods. Present on human skin, hands, the nose and also found in boils, carbuncles and cuts. Contamination on fish occurs from human handling. Toxin survives cooking. Symptoms include vomiting, cramps and diarrhea.

**Steak**: Slices of fish with two parallel surfaces. To cut fish into steaks, use a fillet knife to cut the flesh and a heavier knife (or cleaver) to cut through backbone. Leave the skin on.

**Steaming**: Cooking in steam generated from boiling water.

**Stewing**: A method of cooking based on poaching. Most fish stews use several kinds of fish and vegetables, which simmer in a large amount of liquid.

**Surface slime**: A natural excretion that helps the fish slip easily through the water.

**Tapeworms**: Parasites that live in intestines.

**Tetrodotoxin**: Toxin found in puffer fish, porcupine fish and ocean fish. Toxin is resistant to heat and is present in the reproductive organs, the viscera and the skin. Toxic dinoflagellate algae are suspected as the toxin's origin.

**Trematodes**: Parasites; flatworms.

**Tubes**: Squid mantle processed into tubes.

**Vibrio cholerae**: Bacteria found in estuarine waters particularly in southern coastal states. This bacteria causes cholera. Symptoms include vomiting and dysentery.

**Vibrio parahaemolyticus**: This bacteria is a normal inhabitant of the marine environment, especially in warmer water regions. Frequently contaminates fish, molluscs and crustaceans. Warm temperatures accelerate growth, so the greatest threat is during summer months and improper refrigeration. Symptoms include diarrhea, cramps, vomiting and fever, which usually occur about fifteen hours after ingestion. Bacterium is destroyed by cooking.

**Vinegar:** Fermented cider, malt or wine used in cookery and pickling.

**Viscera:** Internal organs; guts.

**Wings:** Fins of skates and rays.

# Further Reading

1976. *Let's Cook Fish!* Fishery Market Development Series No. 8, USDC/NOAA/NMFS.
1978. *Multilingual Dictionary of Fish and Fish Products*, Fishing News Books Ltd.
1979. *The Good Cook Technique and Recipes: Fish*, Time-Life Books, Inc.
1983. *The Audubon Society Field Guide to North American Fishes, Whales and Dolphins*, Alfred A. Knopf.
1984. *Fishes of the North-Eastern Atlantic and the Mediterranean (FNAM)*, Vol. I, UNESCO Press.
1986. *Fresh Ways With Fish and Shellfish*, Time-Life Books Inc.
1986. *Norwegian Salmon International Chefs' Recipes*, Guldendal Norsk Forlag A/S.
1986. *Quality Handling of Hook-Caught Rockfish*, Marine Advisory Bulletin 20, Alaska Sea Grant College Program, University of Alaska.
1987. *Fishes of the North-Eastern Atlantic and the Mediterranean (FNAM)*, Vol. II, UNESCO Press.
1987. *Fishes of the North-Eastern Atlantic and the Mediterranean (FNAM)*, Vol. III, UNESCO Press.
1988. *World Record Game Fishes*, The International Game Fish Association.
Ade, Robin. 1989. *The Trout and Salmon Handbook*, Facts on File.
Anderson, Robert. 1982. *Florida Saltwater Fish and Fishing, Sharks Included*, National Wildlife Publishing.
Bannerman, A. McK. *Hot Smoking of Fish*, Torry Advisory Note No. 82.

Beck, Bruce. 1989. *The Official Fulton Fish Market Cookbook*, E.P. Dutton.

Bigelow, Henry B., and William C. Schroeder. 1953. *Fishes of the Gulf of Maine*, Museum of Comparative Zoology, Cambridge, MA.

Boehmer, Raquel. 1982. *A Foraging Vacation. Edibles from Maine's Sea and Shore*, Down East Books.

Borgstrom, G., ed. 1961. *Fish as Food*, Vol. I, Academic Press.

Borgstrom, G., ed. 1962. *Fish as Food*, Vol. II, Academic Press.

Borgstrom, G., ed. 1965. *Fish as Food*, Vol. III, Academic Press.

Branyon, Max. 1987. *Florida Freshwater Fishing Guide*, Sentinel Communications Company.

Breder, Jr., Charles M. 1929, 1948. *Marine Fishes of the Atlantic Coast*, G.P. Putnam's Sons.

Burgess, G.H.O., C.L. Cutting, J.A. Lovern and J.J. Waterman. 1965. *Fish Handling and Processing*, Torry Research Station.

Burgess, G.H.O., and A. McK. Bannerman. 1963. *Fish Smoking: A Torry Kiln Operator's Handbook*, Ministry of Agriculture, Fisheries and Food.

Cannon, Ray. 1967. *How to Fish the Pacific Coast*, Lane Publishing Co.

Castro, Jose I. 1983. *The Sharks of North American Waters*, Texas A & M University Press.

Chichester, C.O., and H.D. Graham, eds. 1973. *Microbiological Safety of Fishery Products*, Academic Press.

Daun, Henryk. May, 1979. *Interaction of Wood Smoke Components and Foods*, Food Technology.

Davidson, Alan. 1980, 1988. *North Atlantic Seafoods*, Harper and Row.

Dore, Ian. 1984. *Fresh Seafood: The Commercial Buyer's Guide*, Van Nostrand Reinhold/Osprey Books.

Dore, Ian. 1982. *Frozen Seafood: The Buyer's Handbook*, Van Nostrand Reinhold/Osprey Books.

Dore, Ian. 1989. *The New Frozen Seafood Handbook: A Complete Reference for the Seafood Business*, Van Nostrand Reinhold/Osprey Books.

Dore, Ian, and Claus Frimodt. 1987. *An Illustrated Guide to Shrimp of the World*, Van Nostrand Reinhold/Osprey Books and Scandinavian Fishing Year Book.

Dore, Jennifer, ed. 1988. *Seafood Quiz*, Van Nostrand Reinhold/Osprey Books.

Dudley, Shearon, J.T. Graikoski, H.L. Seagran and Paul Earl. 1970. *Sportsman's Guide to Handling, Smoking and Preserving Great Lakes Coho Salmon*, USDI, FWS, BCF.

Dunfield, R.W. 1986. *The Atlantic Salmon in the History of North America*, UNIPUB.

Erlandson, Keith. 1978. *Home Smoking and Curing*, Barret Jenkins Ltd.

Faria, Susan M. 1984. *The Northeast Seafood Book*, Massachusetts Division of Marine Fisheries.

*Fish and Shellfish Hygiene*, Report of a WHO expert committee convened in cooperation with FAO. Food and Agriculture Organization of the United Nations, Rome. 1974.

Fletcher, Anne M. 1989. *Eat Fish, Live Better*, Harper & Row.

Forbes, R.J. 1955. *Studies of Ancient Technology*, E.J. Brill, Vol. III.

Fougere, H. 1955. *Salting Codfish*, Notes 27, 28, 29, Atlantic Progress Reports No. 52, Fisheries Research Board of Canada.

Freeman, Bruce L., and L.A. Walford. 1974. *Angler's Guide to the United States Atlantic Coast*, Sections I-IV, National Marine Fisheries Service, NOAA, U.S. Department of Commerce, United States Government Printing Office.

Gall, Ken. *Handling Your Catch. A Guide for Saltwater Anglers*, A Cornell Cooperative Extension Publication Information Bulletin 203.

Gorga, Carmine, and Louis J. Ronsivalli. 1988. *Quality Assurance of Seafood*, AVI/Van Nostrand Reinhold.

Gotshall, Daniel W. 1981. *Pacific Coast Inshore Fishes*, Sea Challengers.

Hart, T.J. 1982. *Pacific Fishes of Canada*, Canadian Government Publishing Center.

Hill, Paul J., and A. Mavis. 1975. *The Edible Sea*, A.S. Barnes and Company.

Hoese, H. Dickson, and Richard H. Moore. 1977. *Fishes of the Gulf of Mexico, Texas, Louisiana and Adjacent Waters*, Texas A & M University Press.

Hunter, June. 1988. *Cooking, Pure & Simple*, Kincaid House Publishing.

Iredale, David G., and Roberta K. York. 1983. *A Guide to Handling and Preparing Freshwater Fish*, Department of Fisheries and Oceans, Canada.

Jackson, Roy I., and Dr. William Royce. 1986. *Ocean Forum, an Interpretative History of the International North Pacific Fisheries Commission*, Fishing News Books Ltd.

Jarvis, Norman D. 1950. *Curing of Fishery Products*, Research Report 18, Fish and Wildlife Service, United States Government Printing Office.

Jarvis, Norman D. 1945 (reissued). *Home Preservation of Fishery Products, Salting, Smoking and Other Methods of Curing Fish at Home*, Fishery Leaflet 18, USDI, FWS.

Johnson, James E. 1987. *Protected Fishes of the United States and Canada*, American Fisheries Society.

Joseph, James, Witold Klawe and Pat Murphy. 1988. *Tuna and Billfish, Fish Without a Country*, Inter-American Tropical Tuna Commission.

Kinsella, John E. 1987. *Seafoods and Fish Oils in Human Health and Disease*, Marcel Dekker.

Kramer, Donald E., and Victoria M. O'Connell. 1988. *Guide to Northeast Pacific Rockfishes, Genera Sebastes and Sebastolobus*, Marine Advisory Bulletin #25, University of Alaska.

Krane, Willibald. 1989. *Five-Language Dictionary of Fish, Crustaceans, and Mollusks*, Van Nostrand Reinhold.

Kreuzer, Rudolf. 1971. *Fish Inspection and Quality Control*, Fishing News Books Ltd.

Lamb, Andy, and Phil Edgell. 1986. *Coastal Fishes of the Pacific Northwest*, Harbour Publishing Co.

Lantz, A.W. June, 1964. *A Practical Method for Brining and Smoking Fish*, Trade News, Department of Fisheries of Canada.

Lentz, Ginny. 1983. *Catch of the Day — Southern Seafood Secrets*, Lentz Enterprises.

Linton, E.P., and H.V. French. 1945. *Factors Affecting Deposition of Smoke Constituents on Fish*, Fisheries Research Board of Canada 6 (4).

Manooch, Charles S. 1984. *Fishes of the Southeastern United States*, North Carolina State Museum of Natural History.

Martinson, Linda. 1986. *Simply Salmon: Fresh, Frozen & Canned*, Lance Publications.

McClane, A.J. 1977. *McClane's Encyclopedia of Fish Cookery*, Henry Holt and Company.

McClane, A.J. 1974. *McClane's Field Guide to Freshwater Fishes of North America*, Henry Holt and Company.

McClane, A.J. 1978. *McClane's Field Guide to Saltwater Fishes of North America*, Henry Holt and Company.

McClane, A.J. 1974, 1978. *McClane's New Standard Fishing Encyclopedia*, Henry Holt and Company.

McClane, A.J. 1981. *McClanes's North American Fish Cookery*, Henry Holt and Company.

McClane, A.J., and Keith Gardner. 1984. *McClane's Game Fish of North America*, Times Books.

McKee, Lynne G. June 2, 1961. *An Inexpensive Small Smokehouse From an Old Refrigerator*, Seattle Manuscript Report No. 46.

McLay, R. 1972. *Marinades*, Torry Advisory Note No. 56.

Merritt, J.H. 1969. *Refrigeration on Fishing Vessels*, Fishing News Books.

Morcos, S.R., and W.R. Morcos. 1977. "Diets in Ancient Egypt," *Prog. F. Nutr. Sci.* Vol. 2. Pergamon Press.

Nelson, Joseph S. 1984. *Fishes of the World*, 2nd Edition, John Wiley and Sons.

Nettleton, Joyce A. 1987. *Seafood and Health*, Van Nostrand Reinhold/Osprey Books.

Nettleton, Joyce A. 1985. *Seafood Nutrition*, Van Nostrand Reinhold/Osprey Books.

Ney, Tom. 1989. *The Health-Lover's Guide to Super Seafood*, Rodale Press Food Center.

Nicholson, F.J. *The Freezing Time of Fish*, Torry Advisory Note No. 62.

Nickerson, John T.R., and Louis J. Ronsivalli. 1980. *Elementary Food Science*, AVI/Van Nostrand Reinhold.

Nicolas, John. 1989. *The Complete Cookbook of American Fish and Shellfish*, Van Nostrand Reinhold.

Pearson, John C. (selected and edited). 1972. *The Fish and Fisheries of Colonial North America. A Documentary History of Fishing Resources of the United States and Canada*, National Marine Fisheries Service, NOAA Report No. 72040302.

*Quick Freezing of Fish*, Torry Advisory Note No. 27.

Radcliff, William. 1921. *Fishing from Earliest Times*, E.P. Dutton and Company (out of print).

Robins, C. Richard, Chmn. 1980. *A List of Common and Scientific Names of Fishes from the United States and Canada*, Special Publication No. 12, American Fisheries Society.

Ruiter, A. May, 1979. *Color Smoked Foods*, Food Technology.

Schmidt, R. Marilyn. 1982, 1986. *Seafood Secrets: A Nutritional Guide to Seafood*, Barnegat Light Press.

Schwartz, Frank, J., and J. Tyler. 1972. *Marine Fishes Common to North Carolina*, Division of Commercial and Sport Fisheries, North Carolina Department of Natural and Economic Resources.

Seagran, H.L., J.T. Graikoski and J.A. Emerson. 1970. *Guidelines for the Processing of Hot-Smoked Chub*, Circular 331, USDI, FWS, BCF.

Sidwell, Virginia D. 1981. *Chemical and Nutritional Composition of Finfishes, Whales, Crustaceans, Mollusks, and Their Products*, NOAA Technical Memorandum NMFS F/SEC-11, NMFS/NOAA/USDC.

Sinderman, C.J. 1970. *Principal Diseases of Marine Fish and Shellfish*, Academic Press.

Sink, J.D. May, 1979. *Effects of Smoke Processing on Muscle Food Product Characteristics*, Food Technology.
Smith, Margaret M., and Phillip C. Heemstra. 1986. *Smiths' Sea Fishes*, Springer-Verlag.
*Smoked Fish—Recommended Practice for Retailers*, Torry Advisory Note 14 (revised).
*Smoked White Fish—Recommended Practice for Producers*, Torry Advisory Note No. 9.
Squire, James L, Jr., and S.E. Smith. 1977. *Angler's Guide to the United States Pacific Coast—Marine Fish, Fishing Grounds and Facilities*, NMFS/NOAA/USDC, United States Government Printing Office.
Stansby, Maurice E., and Francis P. Griffiths. *Preparation and Keeping Quality of Lightly Smoked Mackerel*, Research Report No. 6, FWS.
Stroud, G.D. *Rigor in Fish, the Effect on Quality*, Torry Advisory Note No. 36.
Tressler, D.K., and C.F. Evers. *The Freezing Preservation of Foods*, Vol. I, AVI/Van Nostrand Reinhold.
Tressler, Donald K., and James McW. Lemon. 1951. *Marine Products of Commerce*, Reinhold Publishing Corporation.
VanKlaveren, F.W., and R. Legendre, 1958. *A Comparison of Various Salt Cod Products*, Atlantic Progress Reports No. 71, Fisheries Research Board of Canada.
von Brandt, Andres. 1984. *Fish Catching Methods of the World*, Fishing News Books Ltd.
Wagenvoord, James, and Woodman Harris. 1983. *The Complete Seafood Book, An Insider's Guide to Shopping, Preparing, and Enjoying the Harvest of the Sea*, Macmillan Publishing Co.
Walford, Lionel A. 1937, 1965. *Marine Game Fishes of the Pacific Coast from Alaska to the Equator*, Smithsonian Institution Press.
Walford, Lionel A. 1974. *Marine Game Fishes of the Pacific Coast from Alaska to the Equator*, Smithsonian Institution Press.
Walker, Charlotte. 1984. *Fish and Shellfish*, HP Books.

Waterman, J.J. *Fish Smoking: A Dictionary*, Torry Advisory Note No. 83.

Weimer, L., W. Downs, C. Manson and P. Smith. *The ABC's of PCB's*, Public Information Report WIS-SG-76-125, University of Wisconsin Sea Grant Communications Office.

Went, A.E.J. 1980. *Atlantic Salmon: Its Future*, Fishing News Books Ltd.

William, L. September, 1979. *New England Red Tide*, Marine and Coastal Facts No. 6.

Wooten, R., and D.C. Cann. *Round Worms in Fish*, Torry Advisory Note No. 80.

Young, Leo, and N.D. Jarvis. *Smoking Anglerfish, Sea Trout and Spanish Mackerel*, CFR 6, NO. 1.

Yudkin, John. 1985, 1986. *The Penguin Encyclopedia of Nutrition*, Penguin Books.

Zinn, Donald J. 1975. *The Handbook for Beach Strollers from Maine to Cape Hatteras*, The Pequot Press.

# Index

Aerobic, 137
Albacore, 36, 90
Alewives, 32, 36, 50, 107
Allergen, 137
Allergies, 113
Allmouth, 84
Aluminum foil, 26, 28
Amberjack, 90
Ammonia, 72
Anadromous, 137
Anaerobic, 110, 137
Anchorworms, 137
Anglerfish, 84
Anisakiasis, 108, 137
Antigen, 137
Ascorbic acid, 27

Bacalao, 138, 145
Bacteria, 2, 34, 36, 109
Bacteria, spoilage, 16, 49
Bacterium, 111
Baking, 126
Barracuda, 90, 98, 114
Bass, 13, 53, 87, 90, 107, 119
Biotoxins, 97
Bites, 113
Bleeding, 2, 3, 37, 138
Blood, 50

Bluefish, 28, 31, 32, 36, 50, 55, 90, 99, 114, 119
Bocaccio, 90
Bonefish, 67, 90
Bonito, 46, 90, 99
Botulism, 62, 110, 138
Bowfin, 87
Braising, 133, 138
Brevitoxin, 101, 138
Brine, 6, 8, 40, 42, 43, 46, 76
Brine strengths, 41
Brined fish, 63
Broiling, 128
Buffalofish, 69
Bullhead, 50
Burbot, 69, 119
Burlap, 9, 10
Butterfish, 32, 68
Butterfly fillet, 15
Buttons, 138

Calamari. *See* squid
Calcium salts, 39, 50
Canadian cooking rule, 125, 138
Candling, 109, 138
Carp, 64, 69, 119

Catfish, 13, 36, 51, 54, 67, 113, 114, 119
Caviar, 87, 139
Cestodes, 106, 139
Ceviche. *See* seviche
Cheeks, 88
Chilled brine, 8
Chilled seawater, 7, 70, 77
Chilling, 6
Cholera, 111
Cholesterol, 116, 121
Chowder, 88, 139
Chub, 50
Ciguatera, 96, 98, 99, 139
Cisco, 119
Citrus juice, 53
Clams, 77, 96, 100, 111
Cleaning, 3
*Clostridium botulinum*, 109, 110
Cobia, 67, 91
Cod, 13, 32, 41, 68, 69, 87, 91, 105, 108, 119
Cod heads, 88
Cod roe, 68
Cod tongues, 88
Cod worms, 108
Cold smoking, 64
Cookbooks, 136
Cooking fish, 125
Corn syrup, 25
Corvina, 53
Crabs, 113, 121
Croaker, 54, 91, 119
Cunners, 114
Cures, types of, 45
Cusk, 13, 36, 41, 69
Cutting board, 12

Decay, 97
Decomposing, 35
Dehydration, 26, 139
Deterioration in storage, 18
Diphyllobothriasis, 108, 139
Dogfish, 32, 71, 113, 114
Dorsal, 140
Dressing, 140
Drip loss, 140
Drum, 64, 69, 91, 119
Dry-salted fish, 63
Drying, 42, 43, 63
Dysentery, 111

Eels, 64, 114
Enzyme, 2, 58
Evaporation, 26
Eviscerating, 3

Fats, 116, 117, 121, 140
Filefish, 97
Fillets, 33, 44, 53
Fillet, butterfly, 15
Fillet knife, 12
Filleting, 11
Filleting board, 12
Finnan haddie, 64
Fish liver oils, 123
Fish rose, 114, 140
Flake, 71
Flatfish, 105, 119
Flounder, 13, 16, 31, 32, 67, 87, 91
Fluorine, 123
Freezer burn, 26, 141
Freezer wrap, 26
Freezing, 22, 23, 29, 108

Freezing smoked fish, 67
Freshen, 47
Freshness, 22
Frill, 16, 141
Frying, 129
Fugu, 97, 141
Fumet, 141

Gaff, 1, 141
Gar, 87
General Accounting Office, 96
Gills, 5, 141
Glaze, 25, 26, 28
Goosefish, 84
Gravlax, 47, 141
Grayfish, 71
Green-salted, 43, 141
Grouper, 67, 91, 98, 119
Grunt, 91
Gurnards, 86
Gutting, 2, 3

Haddock, 13, 31, 32, 36, 53, 68, 91, 105, 119
Hake, 13, 30, 31, 32, 41, 69, 70, 71
Halibut, 32, 91
Heart disease, 117
Heavy metals, 104, 142
Hepatitis, 112
Herring, 32, 35, 42, 49, 51, 55, 87, 107, 108
Histamines, 97, 100, 142
Hot smoking, 66
Huss, 71
Hydrometer, 40, 142

Ice, 6, 9, 20, 72, 97
Ice crystals, 30
Inspection, mandatory, 96
Interleaved, 27
Iodine, 123

Jacks, 92, 98

Kenching, 42, 45, 142
Kidney, 3, 142
Kokanee, 64
Kosher salt, 39, 50, 61

Labeling, 29
Ladyfish, 92
Lake trout, 53, 87
Leftover fish, 135
Lemon juice, 27
Lingcod, 92, 119
Liquid smoke, 68
Lobster, 113, 121

Mackerel, 13, 28, 32, 36, 42, 44, 46, 48, 50, 51, 55, 67, 76, 92, 99, 108, 113, 119, 120
Magnesium salts, 39, 50
Mahi-mahi, 92, 99, 120
Mantles, 81
Marinating, 49
Marlin, 92
Menhaden, 107
Mercury, 104
Methylmercury, 105
Microwaving, 33, 134
Milts, 87, 142
Minerals, 123
Mollusc, 77, 100, 102, 111, 142

Monkfish, 84, 114
Mother-in-law fish, 84
Mouse unit, 99, 101, 143
Mullet, 92, 120
Mussel, 68, 77, 96, 100, 101, 102, 111
Myxosporidian, 107

N-3s, 121
Nape, 143
Nematodes, 106
Nobbing, 15
Nutrition, fish and, 116

Ocean blowfish, 84
Ocean sunfish, 97
Odor, 18
Omega-3, 89, 116, 119, 122, 143
Onions, 51
Oxidation, 27
Oyster, 77, 96, 100, 111

PCB, 103, 144
PSP, 100, 101, 102
Pacific rockfish, 13
Packages, 25
Packaging, 26, 30
Paddlefish, 87
Pan-ready, 27
Pantothenic acid, 123
Paralytic shellfish poisoning, 100, 143
Parasites, 105, 107, 108, 109, 144
Parrot fish, 98
Pellicle, 63, 144

Pen, 77, 79, 144
Perch, 27, 32, 107, 114, 120
Phosphorus, 123
Pickle curing, 40
Pickling, 35, 49
Pike, 105, 107, 114, 120
Plastic film, 27, 28
Poaching, 132
Pollock, 32, 36, 41, 64, 87, 93, 120
Polychlorinated hydrocarbons, 103
Polyunsaturated fats, 116
Polyunsaturated fatty acids, 117
Pompano, 93, 120
Porcupine fish, 97
Potassium, 123
Processing, 11
Processing at sea, 10
Proteins, 117
Protozoans, 106
Puffer, 87, 93, 144, 145
Puffer-fish poisoning, 97

Quahogs, 100
Quality, 20, 22

Rainbow trout, 120
Rancidity, 26, 27, 145
Rays, 81
Red tide, 100
Refreezing, 33
Refrigerated seawater, 7
Refrigerator, 20, 33
Retardant, 27
Riboflavin, 123

Rigor mortis, 10, 11, 23, 145
Rockfish, 13, 36, 54, 93, 120
Rock salmon, 71
Rock salt, 39
Roe, 87, 120
Roe, cod, 68
Roe, sturgeon, 87

Sablefish, 93, 120
Safety, 96
Sailfish, 93
Salinometer, 40, 42, 53
Salmon, 27, 31, 32, 36, 42, 44, 46, 47, 50, 51, 60, 65, 87, 93, 105, 107, 113, 120
Salmonella, 111, 145
Salt, 9, 39, 44, 46, 50
Salt cod, 35, 36, 145
Saltfish, 45
Salting fatty fish, 44
Salting fish, 35
Sardines, 113
Saturated fat, 116
Sauger, 36, 54, 87
Saurine poisoning, 99
Sautéing, 130
Sawdust, 59
Saxitoxin, 101, 145
Scaling, 11
Scallops, 53, 77, 100
Schillerlocken, 71, 146
Scombroid, 146
Scombroid poisoning, 97, 99
Scombrotoxin, 100
Sculpins, 86, 114
Scup, 32, 114
Sea bass, 36, 98, 120

Sea robins, 86, 114
Sea squabs, 97
Sea trout, 93, 120
Seasonal variations, 22
Seawater, 7
Seawater, chilled, 70, 77
Seaweed, 10
Selenium, 105, 146
Seviche, 53, 146
Shad, 36, 50, 67, 87, 93
Shark, 94, 105, 114, 120
Sheepshead, 69, 120
Shelf life, 67
Shigella, 111
Shrimp, 121
Siphon, 146
Skate, 81, 83, 144
Skinning, 13
Slime, 50, 147
Slush ice, 8
Smelt, 32
Smoked salmon, 64
Smoking, 35, 58, 59, 66, 108
Snapper, 53, 67, 94, 98, 120
Soaking, 42
Sodium, 123
Sodium nitrite, 62
Sole, 94
Soups, 88
Spawning, 23
Spices, 50
Spoilage, 2
Spoilage bacteria, 2, 16, 49
Spot, 94
Squid, 32, 34, 53, 76, 77, 80, 120
*Staphylococcus aureus*, 109, 110
Steaking, 14

Steaks, 33
Steaming, 132
Steelhead, 47
Stewing, 133, 134
Stingrays, 114
Stings, 113
Stir-frying, 131
Stock, 88
Stockfish, 45
Storage, 16
Storage, deterioration in, 18
Storage life, 30
Storage temperature, 30
Storing air-dried fish, 44
Stringers, 6
Striped bass, 32
Suckers, 69, 120
Sugar, 51
Sulfates, 39, 50
Sunfish, pumpkinseed, 120
Surgeon fish, 98
Sweetness, 1
Swordfish, 94, 99, 105, 120

Tapeworm, 107
Tautog, 94
Temperature fluctuations in freezer, 26
Temperature, storage, 30
Tetrodotoxin, 97, 99

Thawing, 33
Thermometers, 60
Thiamine, 123
Tilefish, 94, 120
Trash, 69
Trematodes, 106
Triggerfish, 97
Trout, 42, 51, 53, 67, 87, 120
Tuna, 32, 36, 46, 50, 68, 94, 99, 100, 105, 120

Vinegar, 49, 50, 52, 148
Virus infections, 112
Viscera, 9, 50
Vitamins, 27, 123, 124

Wahoo, 95
Walleye, 13, 36, 53, 54, 64, 87, 105, 114
Water quality, 51
Weakfish, 32, 53, 95
Whelks, 100
Whitefish, 70, 120
Whiting, 13, 31, 32, 69, 70, 120
Whole fish, 33
Wings, 81, 148
Wolffish, 69, 95, 114, 120

Yellowtail, 36, 54, 95